Integrated Farming System: A Strategy for Sustainable Farm Production & Livelihood Security

Editor

Ravikiran Vasant Mane

scitus
academics

Integrated Farming System: A Strategy for Sustainable Farm Production & Livelihood Security

Edited by **Ravikiran Vasant Mane**

ISBN: 978-1-68117-072-5

Library of Congress Control Number: 2015935470

www.scitusacademics.com
Box No. 4766, 616 Corporate Way,
Suite 2, Valley Cottage,
NY 10989

Notice

Printed in United States of America on Acid Free Paper ∞

Contents

Preface

Today food security nutritional security livelihood security, sustainability of productivity and conservation of natural resources are the major issues of agriculture for farm scientists, agriculturists and the policy makers. Being an agrarian country with more than half the population depending on agriculture above issue need to be addressed on priority with due consideration to changing climate. Again majority of holding being managed by small and marginal farmers, it is imperative that any technology generation should primarily aim at these groups and in that farming system being holistic in approach appears one such panacea to address these issues as it takes in to account the potential natural resources, inhabiting human livestock and allied enterprises, and market force. That is why integrated farming system research is the cynosure of all researchers with farm and farm family as the base. This kind of research was initiated during the nineties in TBP command and during last couple of years the University of Agricultural Sciences has initiated model integrated farming units both under rainfed and irrigated ecosystems at all the research stations under RKVY, GOK besides undertaking the task of empowering farmers. The chapters provide a state-of-the-art account of the information available on integrated farming system for farm production and livelihood security.

Editor

Chapter 1

A Basic Strategy to Manage Global Health with Reference to Livestock Production in Asia

David C. Hall and Quynh Ba Le

Department of Ecosystem and Public Health, Faculty of Veterinary Medicine, University of Calgary, Calgary, AB, Canada T2N 2Z6

ABSTRACT

Newly emerging infectious diseases (nEIDs) have increased rapidly presenting alarming challenges to global health. We argue that for effective management of global health a basic strategy should include at least three essential tactical forms: actions of a directly focused nature, institutional coordination, and disciplinary integration in approaches to health management. Each level of action is illustrated with examples from the livestock sector in Asia. No clear example of all three tactical forms in place can be found from developing countries where food security is a significant threat although Vietnam is developing a comprehensive strategy. Finally, an ecosystem health approach to global health management is advocated; such an approach moves away from the traditional single disciplinary approach. Stronger guidance is needed to direct ecohealth research and application in the management of global health.

INTRODUCTION

In the last two decades, as many countries have broadened their network of trading partners, moved quickly to adopt previously unimagined technologies, and hastened the pace of economic migration, it has become increasingly apparent that health too is a global matter that commands a global perspective. The severe acute respiratory syndrome (SARS) pandemic (There was and is still disagreement as to whether or not SARS was a pandemic. By the Center for Disease Control (CDC) standards, SARS was a pandemic (http://www.cdc.gov/ncidod/EID/vol10no11/04-0797_02.htm). The simplest WHO definition of a pandemic at the time was "a worldwide epidemic of a disease". Although the World Health Organization (WHO) did indeed declare that we had entered the early phases of a pandemic, the official declaration of the highest stage of a pandemic, Phase 6, was never made) in 2003 accentuated this point, as did the Influenza Type A/H1N1 outbreaks of 2009. But before one jumps to the erroneous conclusion that globalization is the new demon, we need to recall that there have long been global movements of disease with devastating consequences such as the smallpox epidemic brought upon the Aztecs by invading Spanish Conquistadors in the 1500s [1]. No doubt the infrastructural differences in today's world mean the risk of a pandemic is greater than 500 years ago but so too are our defenses, particularly in terms of scientific technologies, research and knowledge transfer capacity, and institutional coordination. This paper outlines a broad approach to framing a strategy for managing and improving global health. For a global health management strategy to be effective, we argue that tactics must take place in at least three essential forms: actions of a directly focused nature, institutional coordination, and disciplinary integration in approaches to health management. We illustrate each level of action with current examples from the livestock sector in Asia including village level interventions to increase food security in Bangladesh, an example from Vietnam of improved institutional integration and cooperation to control emerging infectious disease, and improved capacity of researchers and other health personnel to adopt an ecosystem approach to health management in Asia.

The complexities of social and ecological interactions are such that no single approach to managing health resources can adequately compensate for inequities that result in poor access to health care at the village level, unsustainable management of resources leading to the emergence of zoonotic disease, or unrealized institutional partnerships to manage regional health risks such as the recent threat of an Influenza Type A/H1N1 pandemic emerging from Southeast Asia and North America. Root problems may show themselves as poverty and hunger as a result of low incomes and lack of market access in remote villages in northern Bangladesh, lack of awareness or concern for environmental contamination resulting from factors (there are of course innumerable factors, but they include attitudes and preferences, sociocultural beliefs, and the cost of change) including low education or inappropriately distributed market power among stakeholders in livestock producing villages in Southeast Asia, or high institutional transaction costs that prevent collaboration or integration of high-level health management strategies in Vietnam. But regardless of the nature and complexity of the root problems, responses to the challenges of addressing health issues of concern to the global population require improved knowledge, expanded capacity, and determined collaboration of individual stakeholders including end users, researchers, and local-to-regional governments and health related institutions.

NEWLY EMERGING INFECTIOUS DISEASES, CHANGES IN THE GLOBAL HUMAN CONDITION, AND THE HEALTH OF SMALL-SCALE LIVESTOCK PRODUCERS

Newly emerging infectious diseases (nEIDs) such as SARS, H5N1, and Nipah virus encephalitis are health issues with global impact. They seem to emerge without apparent warning, they spread with great speed, and they impose severe threats to global health, ranging from human to animal to economic health. Over the last ten years, these and other nEIDs have appeared in either Bangladesh or Vietnam and sometimes both countries, resulting in significant impact on the health and livelihoods of communities through increased morbidity and mortality of humans and livestock. Furthermore, hundreds of millions of dollars of economic damage has been caused from disruptions in local and regional trade, as well as enforced cessation of livestock production. The impact has been most severe among the rural poor—those who are able least to invoke mitigation strategies to the risks posed by global health challenges.

Changes in human socioeconomic conditions include job loss or abandonment leading to poverty and hunger, rapid population growth and uncontrolled urbanization, movements of political, economic, and environmental refugees, exploitative behaviors and policies, and new contacts between wild fauna and humans and their livestock have altered the emergence and transmission of nEIDs [2]. The impact and implications of these changes on the health of small-scale livestock producers can be severe, most particularly where zoonotic diseases are involved. For example, all 93 human cases of highly pathogenic avian influenza (HPAI) in Vietnam, from 2003 to 2005, 42 of which were fatal, had a history of exposure to sick poultry [3].

Contributing to these conditions is the remarkably rapid expansion of livestock ownership in low-income countries. In Vietnam for example, livestock ownership has expanded rapidly in the last decade with more than 80% of small-scale rural households owning some form of livestock [4, 5]. More than 50% of rural livelihoods depend on livestock for nutrition and cash income in both Vietnam [6] and Bangladesh [7], and because the majority of small-scale livestock keepers are women, livestock also generates employment for women and offers them economic decision making power. The importance of this last point is often not fully appreciated with respect to decisions for the welfare of the household. When women who raise livestock are able to decide how and when the income is spent on the household, then they have control over varying and balancing food inputs for children and infants, increasing food security and improving the health of their family.

The important contributions of livestock to health and economic well-being are considered as part of a global veterinary mission, identified by the Steinfeld of the Food and Agricultural Organization (FAO) of the United Nations as a strategy for improving livelihoods in developing countries [8]. Livestock not only provide farmers with a way to increase investment but also promoting security and health of the household. Randolph et al. [9] have outlined the same points, noting the important role of livestock in improving human nutrition and health and reducing and preventing poverty.

Clearly livestock are important in poverty alleviation and health maintenance, but the threat of losing that source of income or possible infection of the livestock keepers themselves increases with each new outbreak of HPAI or other zoonotic disease.

RISKY FARMING PRACTICES WITH FOCUS ON LIVESTOCK PRODUCTION

In the Chars region of north and north-west Bangladesh, a high proportion of villagers are landless, live in extreme poverty, and face annual cycles of flooding resulting in annual hunger and malnutrition. This is a major precipitating factor for health problems among the rural poor, leading to low birth weight, delayed physical and mental development, and susceptibility to infections [10]. The use of livestock and horticulture has been accepted as a suitable intervention by some villagers in the Chars area [11] with initial success. Extending this intervention, integrated agriculture (the incorporation of sustainable crop and livestock practices) in an ecohealth (ecohealth can be defined as the field of study and practice that uses systemic and participatory approaches to understanding and promoting health and well-being in the context of complex socialecological interactions) framework has been introduced as an additional option to address these extreme problems [12]. Preliminary results indicate that household income increased 60–70% when farmers invested in dairy cattle, allowing for economic decisions to be made (and increasing the role of women in making those decisions) regarding purchases of food, clothing, and medicine, and allowing for children to attend local schooling. The result has improved nutrition and economic power of inhabitants, reducing poverty and hunger and increasing child health and employment

(For a more thorough example of integrated health management in an ecohealth framework employed to reduce risk of disease, refer to Joshi and associated papers: Durga Datt Joshi and Minu Sharma. An Urban Ecosystem Health Approach to Make A Cleaner City and Better Health in Kathmandu, Nepal. National Zoonoses and Food Hygiene Research Centre (NZFHRC), Kathmandu, Nepal-Paper presented in World Congress of Public Health, August 18–26, 2006, Brazil.)

Despite this step towards eradicating poverty in the Chars region of Bangladesh, major constraints remain. Productivity of cattle is low due to poor genetics and weak knowledge of management skills, there is limited integration of environmental sustainability options, and coliform bacteria from livestock pose a health hazard to humans who live in close contact with their livestock day and night. The latter is of more than local concern for the reasons described already; recent changes in Escherichia coli have been noted globally, generating apprehension over the possibility of the emergence from the livestock sector of "super-bugs" resistant to available antibiotics and generating world-wide concerns for health of animals and humans. This is also one of the reasons, consistent with observations of the World Health Organization, that poor livestock keepers are more likely to bear the brunt of the impact of zoonotic disease [13].

For now these constraints and concerns in the Chars of the Jamuna River in north central Bangladesh are being addressed locally at the village community level through training programmes, technical interventions, and improved access to health and agricultural services. This then is a good example of the first of the three essential forms of tactics in a global health management strategy (and surely the most common): action at the grass roots level, or perhaps more politely termed directly targeted interventions. Regrettably, the other two essential forms of tactics are in relatively early stages in Bangladesh. Disciplinary integration in approaches to health management is rare, and institutional coordination is the exception rather than the rule, largely due to financial, educational, and to some extent political barriers at the level of institutions. Early progress has been made in the form of a National Livestock Policy that advocates these concepts with an action plan although this policy is yet to be ratified by the Government of Bangladesh [14]. This is an important step towards policy support for an integrated approach that will include improved access to health and agricultural services and targeted technical interventions intended to address poverty while leaving intact options for small-scale backyard livestock keeping.

Without a research environment and appropriate policy that supports wider development and adoption of an integrated ecosystem approach to health management in low income countries where options are limited for rapid mobilization and change, the use of the grass roots level tactic is likely to remain dominant for many years to come. This does not speak well of mitigating the contributing factors and risks of inequities in human health; furthermore, it is a tactic without teeth for managing global health.

Vietnam has its share of directly targeted interventions in the livestock sector as well although as we shall see a more broadly thinking strategy is developing in the livestock sectors of Vietnam to embrace the second and third tactics of institutional coordination and disciplinary integration. Nevertheless, Pfeiffer et al. [15] has noted that in Vietnam "agri-livestock farming systems involving domestic water birds and rice production in river delta areas are important for the maintenance and spread of infection." Restructuring of the livestock sector in Vietnam has already begun, as have the other tactics suggested in this paper.

MOVING FROM SINGLE DISCIPLINARY TO INSTITUTIONALLY COORDINATED APPROACHES

While we argue that directly targeted interventions are necessary, they require the support of other broader tactics to be of relative importance in a wider based approach to managing global health. Livestock policy responses addressing poverty alleviation have tended to focus on intensive short-term interventions targeting single nEIDs (e.g., HPAI control policies in SEAsia), generally within the confines of solitary disciplines rather than long-term sustainable solutions which integrate truly transdisciplinary approaches. In Vietnam, emerging infectious diseases (EIDs) such as avian influenza and the association of livestock with EIDs have generated considerable control-based activities, particularly at the farm level [16].

For highly pathogenic avian influenza (HPAI) and occasionally for pig diseases including swine influenza (H1N1), options have included broad-scale depopulation, targeted culling, movement control, segregation of species of livestock, and vaccination. Knowledge of coordinated large-scale response programmes has increased, and Vietnamese government agencies have developed well-planned and-implemented programmes to respond to nEIDs such as HPAI. Evidence from Vietnam and Thailand indicates that while such programmes may reduce the risk of EIDs after farms have become infected, they do not address pre-infection strategies [17]. Furthermore, they do little to address the longer-term economic and household nutrition needs of poor rural households that are likely to be engaging in small-scale raising of livestock.

It has been recognized by the Government of Vietnam and supporters of pro-poor policies that, in order to maintain access to small-scale livestock which provides income and maintains food security while reducing the precipitating factors of infectious disease, it is necessary to consider restructuring livestock-keeping methods and strategies [18]. Such restructuring methods would not remove livestock but rather recommend methods by which livestock might be raised such that the risks of EIDs are greatly reduced. This could include integrated agriculture rather than focusing on single species of plants and animals, managing water resources to reduce contact of animal species with each other and with humans, developing alternative methods of livestock housing to prevent the transmission of pathogenic organisms, and following simple steps when raising livestock to ensure the health of those livestock and subsequently the health of the families around those livestock. Proposed changes could take place throughout the market chain, not only at the level of the household.

Implementation of some of these changes, which would require transdisciplinary understanding and collaboration, has been tried on a small scale with rather limited success. Visits to participating communes in Vietnam in 2009 by the authors revealed that less than 10–15% of households had adopted more than two of eight main recommendations [19]; no formal study has been published from study of these communes to examine the reasons for low adoption and implementation or the impact on health of women and children. Although the use of a multidisciplinary team is a giant step in a positive direction, we believe that part of the reason for low adoption may well be the focus on a single species outcome rather than an integrated outcome that balances environmental sustainability with community partnership and free choice of economic activities. In other words, results from adopting integrated agriculture and ecosystem approaches to health management have to be driven by basic understanding of the transdisciplinary issues and by free choice to engage in economic activities.

At the institutional level, Vietnam has made significant progress in this area with the development of the Government of Vietnam-United Nations Joint Program (JP) to fight Highly Pathogenic Avian Influenza (HPAI). The JP is a solid example of multi-institutional effort to address nEIDs and national health inequities using the resources of numerous government ministries and international nongovernment agencies. The JP initially addressed the immediate emergency support needed to control the outbreaks of HPAI that started in 2005. The programme has continued to a new phase in which design and implementation has included international coordination/technical assistance, communication, agriculture (livestock production and animal health), and human health involving the Ministry of Agriculture and Rural Development (MARD), the Ministry of Health (MoH), the United Nations Development Program (UNDP), the United Nations Children's Fund (UNICEF), the Food and Agriculture Organization of the UN (FAO), and the World Health Organization (WHO). The overall objective of the program is "to reduce the health risk to humans from avian influenza by controlling the disease at source in domestic poultry, by detecting and responding promptly to human cases, and by preparing for the medical consequences of a human pandemic." More specifically, the program aims at reduced risk of a global pandemic of HPAI emanating from Viet Nam and enhanced national and local capacity to manage outbreaks of diseases of epidemic potential caused by human and animal pathogens.

Some of the key findings from the midterm evaluation of this program were that it has had impact on coordinating within and across the UN agencies and Ministries of the Government of Vietnam, and improved coordination between implementing agencies of the JP has resulted in a more holistic approach to solving a critical health issue for Vietnam and the region. However, the concept of sustainable ecosystem health to prevent emerging infectious disease is present in bits and pieces of activities throughout the JP (e.g., surveillance, training of community animal and human health workers, communication, vaccination, and restructuring) but is not a major pillar of understanding behind many of the activities (and was not intended to be). It was also recommended that although an ecohealth philosophy is evidently emerging in the JP strategy, wider knowledge of an ecohealth approach is needed, particularly at lower administrative levels, in order to appreciate the integrated roles of managing the interfaces of animals, humans, and the environment. Most importantly, this must be directed at nEIDs in general, rather than (e.g.,) HPAI or Influenza Type A/H1N1. Expanding this concept, as a broader theme is current being considered in the development of the new phase of the JP.

BUILDING ECOHEALTH CAPACITY IN ASIA

Finally we examine an example of development of the capacity of researchers and ecohealth professionals to adopt an ecosystem approach to health management in Asia, the third essential form of tactic needed for a global health management strategy to be effective. The nongovernment organization Veterinarians without Borders/Vétérinaires Sans Frontières—Canada (VWB/VSF-Canada) is implementing the Building Ecohealth Capacity in Asia (BECA) project, a three-year initiative to build ecohealth capacity in six Southeast Asian countries: Cambodia, China, Indonesia, Laos, Thailand, and Vietnam [20]. The initiative, funded primarily by the International Development Research Centre (IDRC) and Australian Agency for International Development (AusAID), aims to build knowledge and skills in ecohealth in order to help local ecohealth practitioners identify and reduce factors that contribute to emerging infectious diseases in the region.

A large part of the project focuses on transfer of knowledge to high-level researchers working in lead research institutions in Southeast Asia, as well as government and not-for-profit institutions from the same region working in health and agriculture. Together these partners will help drive the project through network development and through the creation of new ecohealth training and knowledge transfer opportunities. This research project will investigate the processes involved in building the capacity for research and application of ecosystem approaches to health management among researchers, development practitioners, and policy makers in Southeast Asia. The research will also investigate the methodologies and tools which contribute to effective capacity building in ecosystem approaches. Existing regional networks such as the Asian Partnership for Emerging Infectious Disease Research (APEIR) have already begun defining the kinds of research agendas and personal relationships that are necessary to make an ecosystem approach effective.

The BECA research project will enhance opportunities for these and newer regional networks to identify and build capacity through the research activities of this project. The working hypothesis of this project is that bringing actors together from different countries and institutions with an array of experience and expertise in the prevention of EIDs, public health, and health promotion will enable participants to investigate and respond more effectively to complex ecohealth issues, with a particular focus on EIDs. This work will contribute in the longer term to an effective network of expertise in ecosystem approaches to managing health, contributing to ecohealth skills, and knowledge development in the region including input to ecohealth policy formulation. The target audiences of this project include academic researchers, government staff, and extension personnel who are and will be involved in promoting integration and application of ecohealth approaches in activities which directly benefit communities in need.

In brief, there are four main outputs expected from an initiative of this kind:

1 identification and promotion of methodologies for developing and measuring impact of sustainable ecosystem approaches to health;
2 promotion of research and application of trans-boundary sustainable ecosystem approaches to health through informing and influencing local and regional policy formulation;
3 development of a research and training network led by high-level researchers, decision makers, and public health workers in the region to build capacity in ecosystem approaches to health, addressing in particular the control of EIDs in South and Southeast Asia;

4 provision of a process-oriented evaluation and consolidation of adaptive and effective learning and teaching tools for building ecosystem approaches to health.

The project is past the first six months of its activities and has engaged with more than fifty participants through focused discussions and exercises at workshop sessions. While this approach serves to bring similar minded researchers and ecohealth practitioners together, it remains to be seen how well the principles and skills imparted in these training sessions will be incorporated into local and regional activities, programmes, and policies directed at reducing nEIDs. Thus far, at least we can report that several of the network participants have incorporated the knowledge learned from this project into research proposals and at least one policy position working paper addressing, for example, the reemergence of rabies in Bali, Indonesia after decades of being rabies free.

This is not the only project in the region that aims to build capacity in an ecohealth approach to management of health. APEIR is leading research in avian influenza in Southeast Asia that to a large extent relies on integrated approaches to research and application, and IDRC and the International Livestock Research Institute are directing related projects with different outputs. These various initiatives have in common the fact that they all embrace an ecohealth approach in some manner and address similar long-term goals of reduction of poverty and improvement of livelihoods through livestock.

STRONGER EVIDENCE NEEDED FOR ECOHEALTH APPROACHES INTEGRATING LIVESTOCK PRODUCTION

An integrated ecosystem approach to health management is advocated in this paper and in the projects described above as an approach to controlling and preventing nEIDs. Each of the three essential forms of tactics that we have identified would be important for implementing results of ecohealth research, but the groundwork in ecohealth research is still in its formative stages. We are calling for more research directed to ecosystem approaches to health management in order to inform development of sustainable solutions that improve the health and livelihoods of communities. Such approaches examine animal and human health as well as environmental and socioeconomic health and the complex interrelationships between these four dimensions.

Understanding these complexities is critical to exploring the causes of nEIDs and to proposing recommendations for sustainable changes, particularly with respect to methods for raising livestock in a sustainable integrated agricultural system in low income countries such as Bangladesh and Vietnam, barriers to adoption and implementation of such methods, the role of health policy at local, national, and regional levels in promoting and supporting recommended programmes, evaluation of the impact on production, and impact on reducing the precipitating factors of nEIDs. In the particular case of an ecohealth approach that includes livestock within an integrated agriculture system, the goal of improving livelihoods through increased food security will be achieved through healthier livestock and humans, leading to improvements in productivity, increased incomes, and a greater degree of control over economic decision making at the household and community level. Ultimately this means less risk of health inequities in agricultural communities in poor countries and a reduction in global poverty and hunger.

CONCLUSIONS

In this paper, we have presented a broad straightforward approach to framing strategies for managing and improving global health. The three essential forms of tactics that we present—actions of a directly focused nature, institutional coordination, and disciplinary integration in approaches to health management—have been illustrated with examples from the livestock sector in Asia. Evidence from developing countries indicates that this is indeed the general philosophy that has been adopted by governments, research institutions, the nongovernment sector, and other stakeholders working in poverty alleviation through livestock production and integrated agricultural development. It is difficult to find a clear example of all three forms of tactics in use, but the examples presented from Vietnam perhaps present the model in place that comes closest to illustrating this broad approach.

Village level interventions will continue to be important in addressing food security and global health, particularly where communities take a strong role in managing their own resources. Without research, we clearly would be lacking in guidance as to important new directions for knowledge generation and transfer related to correcting health inequities. And finally, the example of the Government of Vietnam-UN Joint Programme to fight Highly Pathogenic Avian Influenza illustrates that cooperation between large complex institutions in addressing a common goal is possible. The latter is not without its challenges, but the JP has been successful enough to warrant its use as a model for a broader One UN concept in which governments, international institutions, and development agencies bridge the sometimes enormous gaps between their institutions to share resources and engage in transdisciplinary problem solving in order to build sustainable solutions to global resources through global health management.

We have also advocated for an ecosystem approach to health management, which is viewed as a more holistic approach to problem solving in global health by examining not just the biological mechanisms of disease but also the complexities of social-ecological interactions. Applied health research, particularly as it relates to zoonotic and nEIDs, does seem to be gravitating in the direction of ecohealth applications. However, clear guidance from ecohealth research is needed to identify primary areas of investigation that will yield sustainable solutions of high impact on poverty, livestock and human health, and environmental management.

REFERENCES

1. J. Diamond, Guns, Germs, and Steel, Norton, London, UK, 1990.
2. R. H. Beach, C. Poulos, and S. K. Pattanayak, "Prevention and control of avian influenza: a view from the farm household," in Proceedings of the CAES-FLP-CATPRN Joint Workshop: Crises in Agriculture and Resource Sectors: Analysis of Policy Responses, Calgary, Canada, October 2006.
3. R. A. Weiss and A. J. McMichael, "Social and environmental risk factors in the emergence of infectious diseases," Nature Medicine, vol. 10, pp. S70–S76, 2004.
4. D. C. Hall, "Competitiveness of livestock farming in Vietnam: constraints and opportunities," inProceedings of the Canadian

Agricultural Economics Society Symposium, Toronto, Canada, May 2009.

5. I. Maltsoglou and G. Rapsomanikis, "The contribution of livestock to household income in vietnam: a household typology based analysis," Pro-Poor Livestock Policy Initiative (PPLPI) Working Paper 21; Food and Agriculture Organization of the United Nations, 2005.
6. D. C. Hall, T. D. Thao, D. V. Minh, and L. V. Lien, "Competitiveness of the livestock sector in Vietnam. A world bank white paper," in Proceedings of the WB-EASRD and FAO-TCIP, Hanoi, Vietnam, December 2006.
7. D. C. Hall and M. G. S. Alam, "Unpublished data from field site visits to four districts in Bangladesh," 2008–2010.
8. H. Steinfeld, "The livestock revolution—a global veterinary mission," Veterinary Parasitology, vol. 125, no. 1-2, pp. 19–41, 2004. · ·
9. T. F. Randolph, E. Schelling, D. Grace et al., "Role of livestock in human nutrition and health for poverty reduction in developing countries," Journal of Animal Science, vol. 85, no. 11, pp. 2788–2800, 2007. · ·
10. T. Ahmed, "Malnutrition in Bangladesh: How Far Are We from Achieving MDG 1?" The International Centre for Diarrhoeal Disease Research, Bangladesh,https://centre.icddrb.org/pub/publication.jsp?classificationID=1&pubID=7904.
11. M. Marks and T. Sultana, "Economic impact of cattle transfers during the CLP›s Asset Transfer Programme (2006–2008)," Chars Livelihoods Programme, U.K. Department for International Development, 2009.
12. M. G. S. Alam, S. Raha, and D. C. Hall, "Project Progress Report: Improving Livelihood through Herd Health Management and Milk Market Access to Poor Farmers Living in Northern Bangladesh," Bangladesh Agricultural University and the Krishi Gobeshona Foundation, 2010.
13. The World Health Organization (WHO), "The control of neglected zoonotic diseases: a route to poverty alleviation," in Proceedings of the WHO/DFID-AHPMeeting, WHO Zoonoses and Veterinary Public Health, Geneva, Switzerland, September 2005.

14. D. C. Hall, M. M. Rahman, N. Ahamed, et al., "Livestock Policy and Action Plan," Department of Livestock Services, Ministry of Fisheries and Livestock, Government of Bangladesh, Dhaka, 2005.
15. D. U. Pfeiffer, P. Q. Minh, V. Martin, M. Epprecht, and M. J. Otte, "An analysis of the spatial and temporal patterns of highly pathogenic avian influenza occurrence in Vietnam using national surveillance data," The Veterinary Journal, vol. 174, no. 2, pp. 302–309, 2007. · ·
16. T. Sproul, D. Zilberman, J. Ifft, D. Roland-Holst, and J. Otte, "Economics of avian flu policy," HPAI Research Net Research Brief 18, 2009.
17. J. Henning, D. U. Pfeiffer, and L. T. Vu, "Risk factors and characteristics of H5N1 highly pathogenic avian influenza (HPAI) post-vaccination outbreaks," Veterinary Research, vol. 40, no. 3, pp. 1–12, 2009. · ·
18. Government of Vietnam, "The Government of Vietnam—United Nations Joint Program (JP) to fight Highly Pathogenic Avian Influenza (HPAI)," Hanoi, 2009.
19. D. C. Hall and L. B. Quynh, "Midterm evaluation of the government of Vietnam—UN Joint program on avian influenza," Final Report, Agrifood Consulting International, Bethesda, Md, US, 2009.
20. D. C. Hall, "Building capacity for research and practice in ecosystem approaches to health in Southeast Asia," 1st Interim Technical Report, International Development Research Centre, Ottawa, Canada, 2010.

Chapter 2

Pastoral livelihoods of the Fulani on the Jos Plateau of Nigeria

Ayodele O Majekodunmi[1], Akinyemi Fajinmi[2], Charles Dongkum[2], Alexandra P M Shaw[1], and Susan C Welburn[1]

[1]Division of Pathway Medicine and Centre for Infectious Diseases, School of Biomedical Sciences, College of Medicine and Veterinary Medicine, The University of Edinburgh, Chancellor's Building, 49 Little France Crescent, Edinburgh EH16 4SB, UK

[2]Nigerian Institute for Trypanosomiasis Research, P.M.B. 1303, Vom, Plateau State, Nigeria

ABSTRACT

The Jos Plateau is an important cattle-producing area in Nigeria, with a high concentration of pastoral Fulani. In recent decades, pastoralist studies have focused on arid and semi-arid areas, with few based in the sub-humid zones, such as the Jos Plateau. There have been significant socio-economic and agricultural changes in this area over time which makes it necessary to assess current pastoral livelihood strategies. A pastoral livelihood analysis of 66 individual households/herds across 30 villages was conducted, using participatory methods and structured questionnaires. The majority of pastoral households are in the middle or better off wealth categories and only a few are poor in terms of livestock holdings. However, the majority are poor in terms of land holdings. Livestock sales remain the major source of income, but only 30% of households rely on this as their sole source of income. Sale of crops, milk and off-farm activities contributed income in other households. Comparison with previous livestock productivity levels on the Jos Plateau shows that natural resource conflict has increased, with negative effects on productivity, although vulnerability to shocks has decreased over the years. Overall productivity has decreased and calving rates were particularly low. The pastoral Fulani community is relatively well-off, compared to similar populations across West Africa. However, the high proportion of wealth concentrated in livestock and large numbers of land-poor households indicate that there is still a degree of vulnerability to the risks of drought/dry season, disease and conflict in pastoral livelihoods in the area.

BACKGROUND

Small-scale livestock production is a major contributor to the economies of sub-Saharan countries, especially for the rural populations. It also plays an essential role in food production, and the protein provided by livestock products is an important component of this. There are 120 million pastoralists worldwide, 50 million of these in sub-Saharan Africa where they constitute 12% of the rural population (Rass [2006]). The Fulani are the primary pastoral group in Nigeria with a population of approximately 15.3 million. The Fulani first settled on the Jos Plateau around 1910, attracted by the low levels of disease and abundant grasslands, and are present in large numbers till today (Blench [2004]).

The ability of pastoralists to fulfil their role as food producers is directly dependent on the viability of their livelihoods and the factors affecting their livelihood strategies. The livelihoods of pastoral people depend on three things: first, access to assets such as land, livestock, pasture, water, animal health services, community networks, markets, credit and education; second, the environment in which these assets are combined for production and consumption, specifically the political, organisational and institutional infrastructure within which they operate, which affects their ability to use these assets to achieve positive livelihood outcomes; and third, the dynamic context of risks (human and animal disease, marginalisation, drought, civil conflict and competition for natural resources) and seasonal and continuous trends that affect assets and their environment and determine the vulnerability of livelihoods (Rass [2006]; Eldis.org [2010]). Vulnerability is a measure of the lack of resilience to risks and seasonal and long-term trends. Therefore, it is not only an important feature of poverty but also a potential cause. Vulnerability is closely linked to asset ownership - the more assets people have, the less vulnerable they are. The greater the erosion of assets, the greater the level of vulnerability - poor herders are more likely to end up with an unviable herd size than wealthier herders, even when they lose the same relative amount of livestock (Rass [2006]).

The literature on pastoral livelihoods and customs focuses on pastoralists located in the arid/semi-arid zones but less emphasis on those located in the sub-humid zones such as the Fulani of the Jos Plateau (Catley [2000]). Recent changes in population, land use, socio-economic climate and tsetse and trypanosomiasis risk on the Jos Plateau have had significant effects on pastoral livelihoods on the Plateau. Little information is available on the current state of pastoral livelihoods or the drivers for the epidemiology of trypanosomiasis in this area. Therefore, a longitudinal research project was carried out between 2007 and 2008 to provide a comprehensive summary of the epidemiology of bovine trypanosomiasis on the Jos Plateau; determine which social, economic, ecological and cultural factors influence animal health and disease control by herders; provide an overview of current pastoral livelihoods including quantitative analysis of animal productivity and profitability to herders and examine the effects of recent changes in socio-economic and security factors on pastoral livelihood strategies. Results on the epidemiology of trypanosomiasis and social factors affecting it have been published in Majekodunmi et al. ([2013a], [b]). This paper is concerned with livestock productivity and the current state of pastoral livelihoods on the Jos Plateau.

STUDY AREA

The Jos Plateau (Figure 1) in north central Nigeria covers an area of 8,000 km^2 at an average altitude of 1,280 m. It consists of younger granite of volcanic origin, and the terrain is characterised by numerous rocky, flat-topped hills and crater lakes. The Jos Plateau, formerly open savannah woodland but now mostly grassland, is a highly populated, intensively cultivated area. It was historically free of tsetse flies and trypanosomiasis. The absence of trypanosomiasis and abundant pasture attracted large numbers of cattle-keeping pastoralists. The area now holds over a million cattle and plays an important role in the national/regional cattle industry. However, beginning in 1982, there have been increasing reports of tsetse flies and African animal trypanosomiasis (AAT) on the Plateau (Majekodunmi et al. [2013a]).

Figure 1: Jos Plateau.

The Plateau shows high ethno-linguistic diversity, without any of the large-scale political units that characterise Yoruba- or Hausaland. In pre-colonial times, the Plateau was populated by a great variety of small ethnic groups living in hamlets with a complex clan organisation and ritual kingship systems speaking an assortment of 'Plateau' languages and a few Chadic languages related to Hausa. Therefore, no single language or people were dominant, although the largest ethnic groups were (and still are) the Birom, Ngas and Tarok (Gunn [1953]; Blench [2004]). These tribes lived mostly in the natural fortresses provided by the hills, and the central plain of the Plateau was largely uninhabited. However, after the advent of the British at the turn of the century, hill tribes were resettled on the central plain. Soon after, the establishment of the tin mining industry attracted workers and traders from across the country. In the early nineteenth century, low human populations, absence of tsetse and mosquitoes and unlimited grassland made the Jos Plateau one of the most attractive environments for cattle in northern Nigeria. These favourable conditions drew Fulani pastoralists from the semi-arid regions who established themselves across the Plateau. From the 1970s onwards, the Fulani also settled on the lowland plains. Currently, the Jos Plateau is a highly populated cosmopolitan area, with inhabitants from across the country with a significant Muslim/Hausa presence (Fricke [1979]; Awogbade [1983]; Blench [2004]). Rural areas are mostly inhabited by farmers of the original Plateau tribes with a significant minority of settled Fulani herders.

Jos: Conflict and Civil Strife

The population of Jos city is composed of 'indigenes' from Plateau tribes (mostly Christians of the Birom, Anaguta and Afizere tribes), and Hausa-Fulani Muslim 'settlers' comprise a significant minority along with other settlers from across the country (Blench [2003]). The conflict in Jos stems from competition between Muslims and Christians, 'settlers' and 'indigenes', respectively, for political power and representation in government. The Plateau tribes, who are mostly Christians, claim 'indigene' status and ownership of Jos city as the original inhabitants of the land. The Jasawa (Hausa Muslims of Jos) are currently classed as settlers which places them at a disadvantage in terms of social, political and economic benefits such as scholarships, places in state schools and employment in the civil service. The Hausa population in Jos wish to claim indigeneity in Jos, based on their long-standing settlement (fourth or fifth generation in some cases), and demand adequate representation in government and politics. This is vehemently opposed by the 'indigenes' who wish to keep these resources for themselves and feel that the settlers already have enough of an advantage over them in terms of numbers and affluence.

This tension has led to recurring violence in Jos North Local Government Area (LGA), since 1994. In August of 2001, a Hausa Muslim was appointed as the poverty eradication coordinator in the area. Tension escalated until September 2001 when it boiled over into riots that engulfed the whole city and was replayed in rural areas. In one week, over a thousand people were killed and tens of thousands of Muslims (including pastoralists) fled, never to return (Bawa and Nwogwu [2002]). These were reported as religious riots, and those fomenting violence on either side were quick to incite their followers on religious grounds (Human Rights Watch [2001]). There had been relative peace in rural areas of the high Plateau since the violence of 2001 and 2002. In November of 2008, the violence was repeated in disputes over the local government elections in Jos North. More than 780 people were killed in the ensuing riots (Higazi [2011]). But unlike in 2001, the 2008 riot did not lead to violence in the countryside. There were also outbreaks of violence in November 2009 to January 2010. In December 2010 to January 2011, several churches were bombed during midnight mass on Christmas Eve and widespread riots followed. This time, the violence did spread into the rural areas on the Plateau (Higazi [2011]; Krause [2011]). Most of this violence was directed against Muslims - Hausa and Kanuri residents in historic tin mining settlements and adjacent Fulani agro-pastoralist communities - and most were driven out at this time. After a few months, the Fulani regrouped in certain areas and there were repeated attacks on Birom villages. Those affected on both sides lost their homes, farms, livestock and other assets. This current phase of rural violence began in January 2010 and has continued intermittently since then in three of the 17 local government areas in Plateau State: Barkin Ladi, Riyom and Jos South. The loss of life during this period has been even greater than that in 2001 to 2002, and the conflict has become protracted in these areas despite military interventions (Higazi [2013]).

Off the Jos Plateau, the other major centres of conflict in Plateau State are Langtang and Wase local government areas on the lowland plains some 200 km south-east of Jos. The Tarok people are the indigenous tribe in this area, but there are substantial settlements of Hausa farmers and Fulani pastoralists (Blench [2004]). The Tarok have maintained good relations with these settlers for a long time and are now themselves substantial cattle owners, often as a result of sending their sons to be trained in herding by the Fulani. The Tarok, moreover, have a long tradition of military service, and many of their leaders are ex-generals. Apparently, a fight broke out in Yelwa at the end of June 2002 between Christian and Muslim residents resulting in the burning of churches. It quickly spread to other communities in the area and degenerated into protracted guerrilla warfare with each side conducting armed reprisal raids on the other until the end of 2004 (Blench [2004]). Cattle rustling by the Tarok was rife during this period, and of the 1.2 million cattle recorded in Wase LGA, an estimated 500,000 were lost to rustlers (Higazi [2008]). In May 2009, state security forces stormed five villages in Wase LGA and hundreds of Fulani were forcibly loaded into trucks and driven 'back' to Bauchi, Gombe, Katsina and Jigawa states. Insecurity has discouraged both farming and herding in many areas with serious effects on livelihoods and food supply (Blench[2004]).

METHODS

To determine trypanosomiasis prevalence at 95% confidence, a longitudinal two-stage stratified cluster sampling design was used where randomly selected cattle were the ultimate sampling unit and villages were clusters. Thirty individual villages were chosen using a 15 km × 20 km grid applied over the study area as shown in Figure 2. A structured questionnaire, incorporating various participatory rural appraisal (PRA) techniques, was administered to livestock owners who participated in the trypanosomiasis survey. This allowed us to gather data for quantitative analysis of animal productivity and profitability to pastoralists; determine which social, economic, ecological and cultural factors influence animal health and disease control by herders and investigate knowledge, attitude and practices of animal husbandry and disease control.

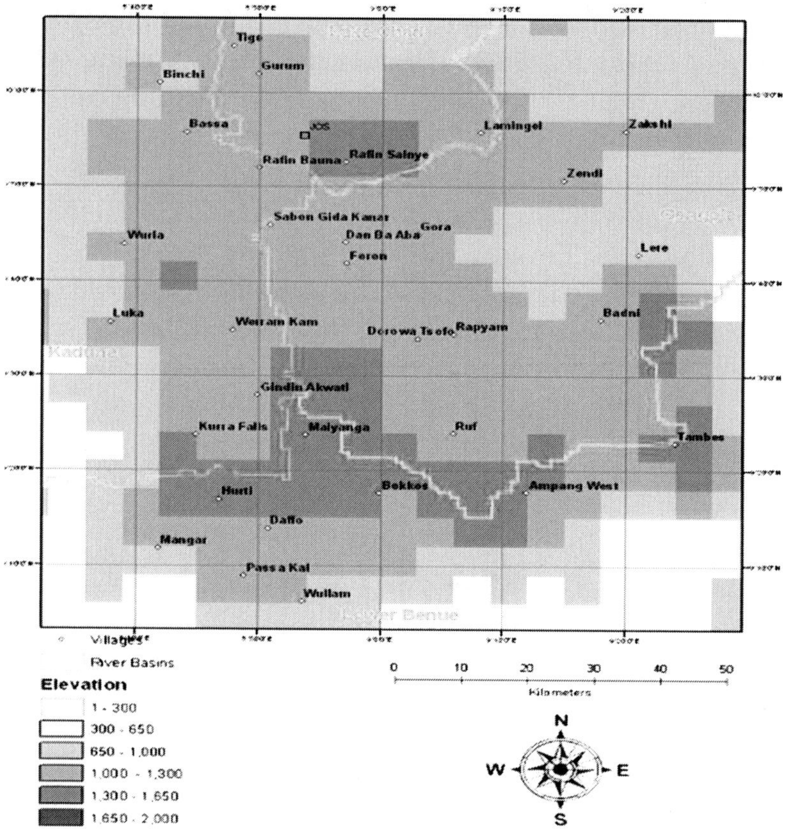

Figure 2: Study area showing selected villages.

Structured questionnaires are a standard method for collecting survey data. However, large-scale surveys and long questionnaires can be biased, unreliable and difficult to administer and analyse. PRA uses a range of qualitative sample survey methods to gather information by allowing local people to share and analyse their knowledge and experiences and to make plans based on this information (Chambers [1994]). It is quicker and more cost-effective than long-term studies and uses a wide range of multidisciplinary information. It enables on-the-spot assessment and direct response to actual village-level problems. It also facilitates collection of standardised field data for use in more sophisticated databases. These qualitative methods may be used on their own or alongside more formal quantitative methodologies. Structured questionnaires incorporating several PRA techniques were administered to livestock owners in individual interviews and focus group discussions focusing on animal husbandry, pastoral livelihoods and the factors affecting them. Household/herd information was collected from 66 individuals for land ownership, mortality, herd dynamics and productivity to allow quantitative analysis of the data.

Households were assigned to wealth groups according to the categories devised by Famine Early Warning Systems Network for pastoral households (FEWS [2004]) as shown in Table 1.

Table 1: Pastoral household wealth grouping criteria (FEWS NET 2004/5)

Wealth group	Household size	Livestock	Area planted
Poor	3 to 5	5 to 10 cows, 10 to 26 shoats	1 to 2 ha with household labour
Middle	7 to 8	15 to 50 cows, 35 to 100 shoats	3 ha with household labour
Better off	8 to 13	50+ cattle, 100+ shoats	5 ha with household labour and hired hands

Majekodunmi *et al.*

Majekodunmi *et al. Pastoralism* 2014 4:20 doi:10.1186/s13570-014-0020-7

RESULTS

Source of Income

Livestock sale was the most important source of cash income amongst households, accounting for 52% of pastoral income across the study area as shown in Figure 3, ahead of crops (22%) and milk (12%). In terms of individual household income diversity, 30% of households relied on livestock as their sole source of income as shown in Figure 4. Forty percent supplemented the income from livestock with crops and 15% with milk. Fifteen percent gained additional income from off-farm activities, mostly mining and sale of firewood.

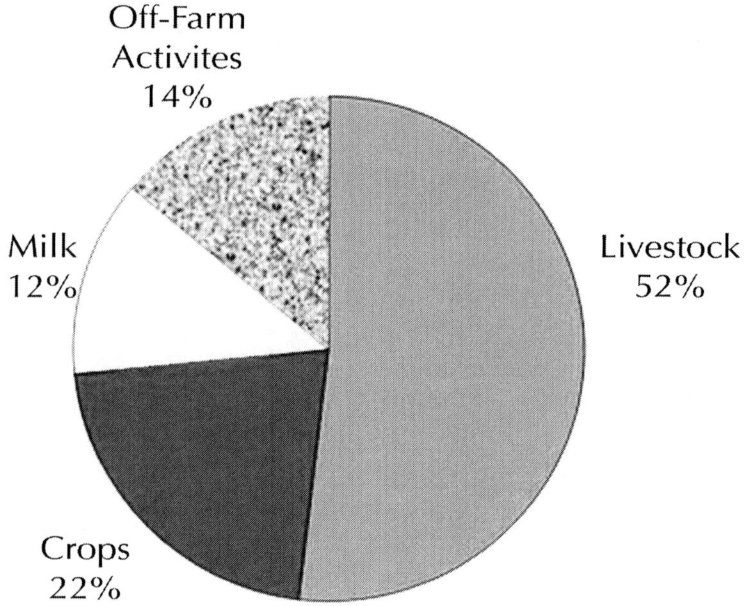

Figure 3: Sources of pastoral cash revenue across the study area.

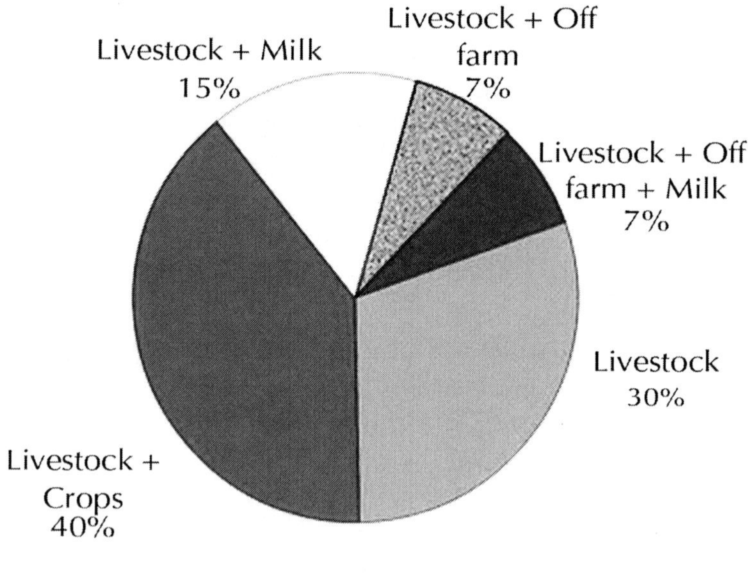

Figure 4: Household income diversity.

Land Holdings

Land ownership was universal, with all households owning a piece of land. The majority of households (61%) owned 1 to 10 acres (Figure 5). The average amount of land owned was 17.9 acres, of which 10.9 were used for cropping while 7 were used for livestock. This illustrates the general trend of land use primarily for cropping purposes. Livestock were generally grazed on commons/fallow/unused land and only tethered at night on land owned by the household.

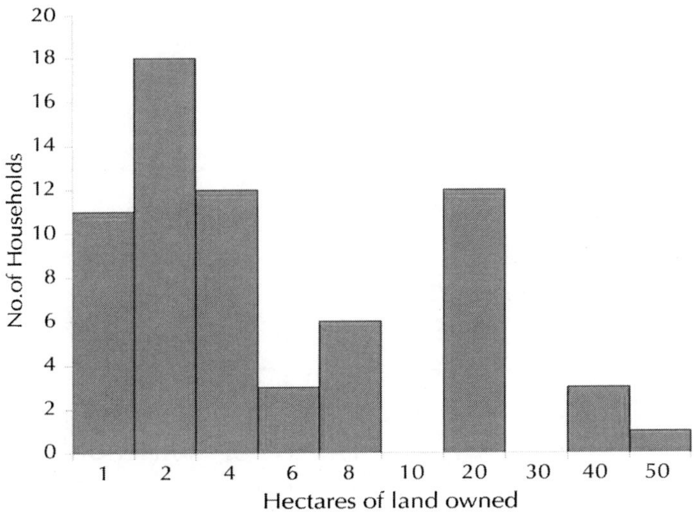

Figure 5: Land ownership.

Crop Farming

All households were engaged in crop farming. One hundred percent of households grew maize, 91.7% grew vegetables (potatoes, lettuce, carrots, peas, cabbages, etc.) and 76.4% grew millet (Figure 6). All recipients grew more than one crop and the average number of crops grown was four. All but three of the 30 households used manure from their own cows (91.9%) and commercial fertilisers to enhance crop yield and quality, but less than half (40.5%) used herbicides.

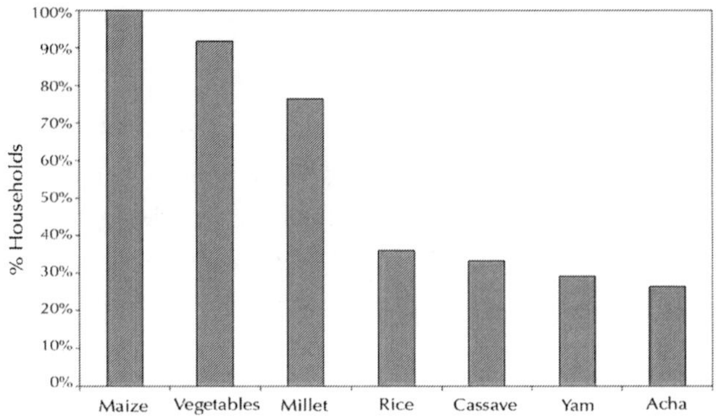

Figure 6: Crops grown by households.

Cattle Holdings

Herd size was very variable, ranging from 7 cattle to 5,015 cattle. The herd size distribution was bimodal, with peaks in the range 10 to 20 cattle and 50 to 100 cattle (Figure 7). The average herd size was 253 cattle. However, when the four outlier herds with >1,000 animals were removed from the sample, the average herd size fell to 78 cattle. All households owned livestock other than cattle, with most of them keeping small ruminants and chickens for household consumption and/or sale and dogs for hunting and/or security.

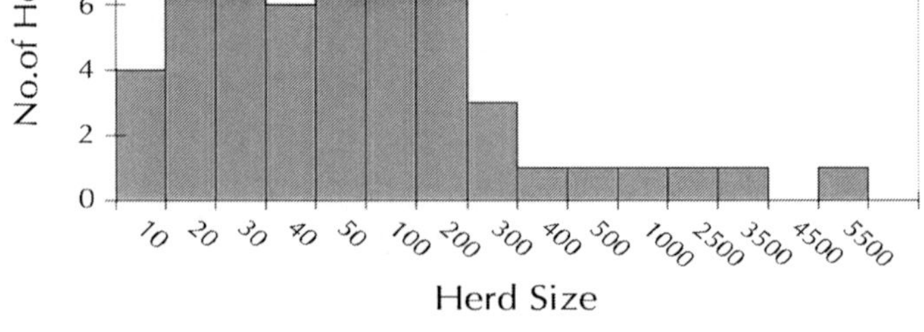

Figure 7: Cattle holdings.

Herd Composition

The age and sex of the sampled animals were recorded. Since the sampled animals were a good representative sample of the total herd, this gives an indication of the total herd composition. The sampled animals consisted of 74.1% (confidence interval: 73.1% – 75.1%) females and 25.9% (24.9% – 26.9%) males.

When considered by age group, 8.3% (7.5% – 8.7%) of the herd were young calves below 6 months old, of which 3.8% (3.4% – 4.2%) were males and 4.5% (4.0% – 4.8%) females. The overlapping confidence intervals indicate that there is no significant difference in sex ratio. Juveniles aged between 6 months and 2 years of age accounted for 25.3% (24.2% – 26.2%): 12.5% (11.7% – 13.3%) males and 12.8% (12.0% – 13.6%) females. The overlapping confidence intervals indicate that there is no significant difference in sex ratio. Those above 2 years old were 66.4% (65.4% – 67.6%): 9.6% (8.9% – 10.3%) males and 56.8% (55.6% – 57.9%) females (Figure 8). There is no overlap in confidence intervals, indicating that there is a significant difference in sex ratio in this age group. There is no overlap in confidence interval between the three age groups, indicating a significant difference in age group composition.

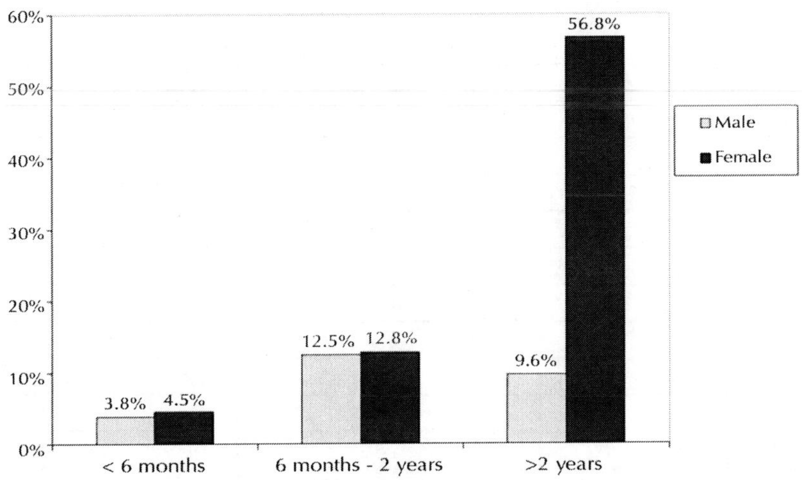

Figure 8: Herd composition.

Herd Productivity

Information on herd size, births, deaths, sales, purchases and prices was collected from individual members of focus group discussions using PRA techniques embedded in a structured questionnaire. Mean values per herd/household are presented in Table 2. The average pastoral cattle herd contained 253 cattle, consisting of 21 calves below 6 months, 64 juveniles between 6 months and 2 years, 143 cows and 25 bulls. In the course of one year, 15 calves would be born (adding 5.8% – the herd) and 2 of these would die (12.7% calf mortality rate). Sales would consist of eight adult cattle (3.2% of herd) sold for an average price of USD 681 each, and three yearlings (1.0% of herd) would be purchased for an average price of USD 309 each. Nine cattle would die (3.5% of herd) and one (0.4% of herd) would be slaughtered for ceremonial purposes. Natural herd growth (births to deaths) would be four cattle (1.6% of herd). Actual herd growth is obtained by adjusting for net offtake (sales + slaughters) that is for eight cattle (3.6% of herd) offtaken less three (1.0% of the herd) purchased, so six (2.6% of the herd). In this description, cattle numbers were rounded to whole numbers, but percentages reflect the fractions. The actual herd growth, recorded for the whole cattle population, was negative, at −1.2% (Table 2).

Table 2: Mean productivity parameters per household herd (with confidence intervals)

	1975	2008
Starting stock	126 (19 – 232) cattle	253 (62 – 443) cattle
Closing stock	141 (16 – 267) cattle	249 (60 – 440) cattle
% Change in herd	+7.8% (−17.0% – +25.6%)	−1.2% (−2.8% – +0.3%)
Births	28.5% (4.0% – 53.0%)	5.8% (4.0% – 7.6%)
Deaths	8.0% (0% – 55.3%)	3.5% (2.1% – 4.9%)
Sales	10.4% (5.3% – 52.6%)	3.2% (2.2% – 4.3%)
Purchases	1.2% (0% – 2.3%)	1.0% (0.4% – 1.7%)
Price bought	$769.5 (115.2 – 1,423.9)	$308.9 (265.0 – 352.8)
Price sold	$559.1 (188.3 – 923.00)	$680.9 (609.2 – 752.5)
Natural herd growth	14.7% (8.2% – 21.2%)	1.6% (0.4% – 2.8%)
Actual offtake	11.7% (2.2% – 21.3%)	3.6% (2.4% – 4.7%)

Majekodunmi *et al.*

Majekodunmi *et al.* *Pastoralism* 2014 4:20 doi:10.1186/s13570-014-0020-7

These herd productivity results were compared to those collected in 1975 by Pullan and Grindle ([1980]) to assess changes over time as shown in Table 2. Prices from 1975 were adjusted for inflation to generate 2008 equivalents (Central Bank of Nigeria [2007]). There were no statistically significant differences between the two data sets.

Wealth Grouping

Analysis of households by livestock holdings revealed that 43.9% of households were 'better off', 50% were in the 'middle' group and only 6.1% were 'poor' (Figure 9). Analysis by land holdings showed that 37.9% of households were 'better off', 7.6% were in the 'middle' group and 54.5% were 'poor' (Figure 10).

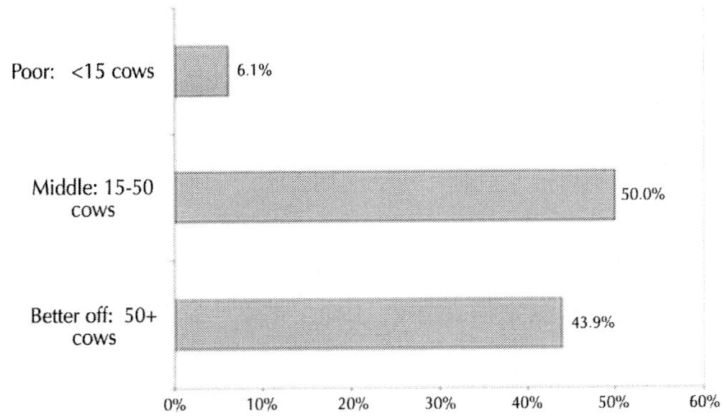

Figure 9: Wealth groups by cattle.

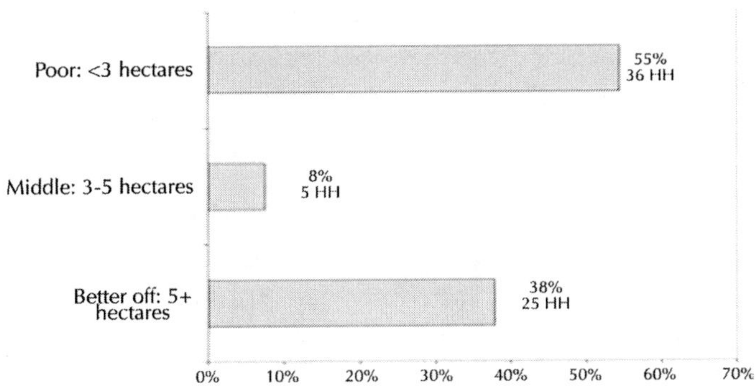

Figure 10: Wealth groups by land.

Conflict, Competition and Migration

Results showed that of the 66 respondents, 44 (67%) practised transhumance: 4 (6%) only during the dry season, 10 (15%) only during the wet season and 30 (46%) during both wet and dry seasons. Twenty-two households did not practise transhumance. The reasons given by respondents for their migration practices in the dry and wet seasons are shown in Figure 11. Figure 12 illustrates the seasonal calendar and migration periods.

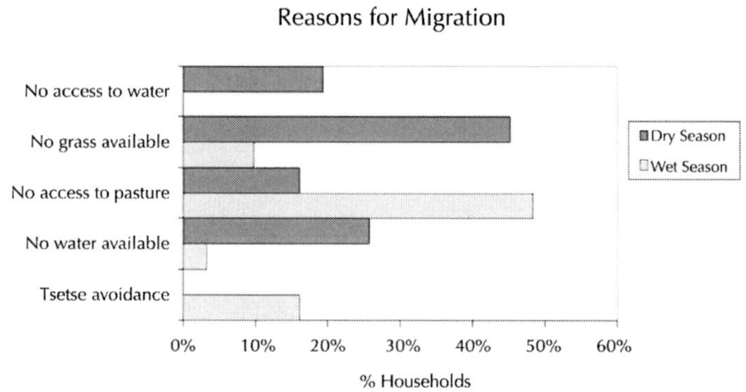

Figure 11: Reasons for seasonal migration.

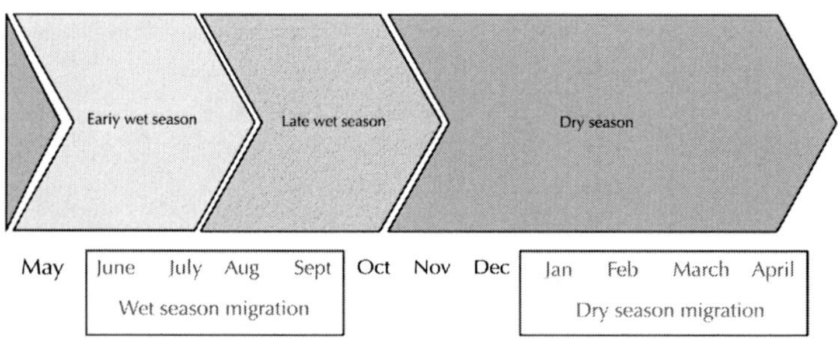

Figure 12: Seasonal calendar with migration periods. Dry season migration takes place between December and April and on average lasts three months. Rainy season migration takes place between April and October and on average lasts two months. Dry season destinations are more varied, half of them being on the eastern side of the Jos Plateau. The destinations vary in distance from a one-day to a three-week trek, but the average distance is a five-day trek.

Dry Season Migration is Practised by Herders in Search of Adequate Pasture and Water

Climatic conditions mean that there is very little grass or water available during the dry season. Pastoral access to these scant resources is further restricted by dry season farming activities practised by the indigenous populations that are common on the Jos Plateau and take priority over cattle grazing. This means that even when grazing and/or water is available, cattle may not be able to access it and herders move them to areas with less land pressure.

Rainy Season Migration is Practised for

- *Accessible grazing and water*: Pasture and water are widely available in the wet season, but uptake of more commercial and intensive farming methods and a higher proportion of cash crops mean that arable farmers increasingly demand more land and water for irrigation, so herders are marginalised.
- *Avoidance of farmers' crops*: Increasing populations and land pressure have intensified farmer/pastoralist conflicts and increased the frequency of animals trampling crops. In many areas, these issues have been resolved simply by asking pastoralists to remove their cows from the village from the time the crops appear above ground until harvest.
- *Avoidance of tsetse and/or biting flies*: In areas where flies are a severe nuisance, cattle are moved to more favourable areas of the Plateau or to nearby lowlands.

DISCUSSION

Sources of Household Income

Income diversity is significant, but 30% of households are still reliant on a single source - livestock sales. Only 16% engaged in off-farm activities, mostly labour-intensive mining/sale of firewood rather than investment/trade. This is an indicator that pastoral livelihoods in this area retain a certain degree of vulnerability as higher income diversity and diversification into trade both reduce vulnerability to risks or shocks.

On the Jos Plateau, the proportion of better off herders is higher than average figures for pastoralists (10% poor, 10% better off and 80% in the middle group; FEWS [2004]). This indicates that their relative 'wealth' and local opportunities (land acquisition, proximity to towns, good road networks, good market access, high demand and high price for vegetables) have encouraged voluntary diversification into growing cash crops in 39% of households for whom these crops bring in 25% of their cash income. The fact that the remaining 61% of households were unable to take advantage of these conditions to diversify into growing cash crops is supported by the results of the wealth grouping by land holdings that show that 54.5% of households were 'land poor'. Those in this category lack the opportunity of land acquisition and are therefore unable to diversify into cropping.

The other main source of income for income-diverse households was the sale of milk. Milk is the main subsistence product of cattle. There is therefore constant competition for milk, between human consumption, calf consumption and sale which may be responsible for the low percentage of households (28%) gaining income from the sale of milk. There is evidence elsewhere that only better off pastoral households consume milk while poorer households prefer to sell it and buy grain to eat (Niamir [1982]). The low number of households gaining income from milk is also as expected based on the wealth grouping results - a higher number of better off households will reserve their milk for consumption by humans and calves while only the few poorer households will sell theirs.

Another important obstacle to diversification into the dairy trade is market exclusion, for which there are two reasons. Firstly, the main potential for pastoral dairying is an informal market serving high urban demand; however, pastoralists in remote areas simply do not have access to this market. Secondly, the entire trade in pastoral dairy products faces stiff competition from imported powdered milk and intensive, peri-urban dairy farms. This competition tends to reduce demand for pastoral dairy products that are viewed as unhygienic in comparison. The lack of demand is a very important constraint to market access as well as to any prospects for its improvement. Demand for pastoral dairy products has been identified as the limiting factor to market participation and as an important pre-requisite for interventions designed to increase market access for pastoralists, e.g. establishment of milk collection centres and improved road network (Michael [1987];

Little [1989]). Since income from the pastoral dairy trade is controlled by women, this also has serious implications for gender empowerment and household economics.

Wealth Grouping

The results showed a discrepancy between wealth grouping by land and wealth grouping by livestock holdings. Grouping by livestock holdings showed that 43.9% of households were 'better off', 50% were in the 'middle' group and only 6.1% were 'poor'. But grouping by land holdings showed that 37.9% of households were 'better off', 7.6% were in the 'middle' group and 54.5% were 'poor'. This shows that although few of the Fulani on the Jos Plateau are poor in terms of livestock, many are still land-poor. There is a continuous distribution amongst wealth groups by livestock holdings, but the distribution of wealth groups by land was bimodal: households owned either a lot of land or very little, with few households in the middle category. This pattern of land ownership may be due to the duration of settlement of different groups of Fulani. Fulani that have been settled for longer are more likely to be granted permission to acquire land, but they are also more likely to be attracted by the advantages of mixed farming than those who have only recently settled in an area. This distribution may also be linked to the amount of labour required to till large pieces of land once acquired. Larger, wealthier households are more likely to have spare hands for farm labour or spare cash to hire labourers. The majority of Fulani with large land holdings contract the tilling and weeding to labourers from the indigenous population, but family labour was used for harvesting. There were no reports of animal traction used by either the Fulani or indigenes for tilling their land.

The results of wealth grouping by livestock holdings show that few herders are in the poor (6.1%) group while the majority are in the middle (50%) and better off groups (43.9%) on the Jos Plateau compared to average figures from FEWS NET wealth analysis which showed 10% pastoral households to be poor, 10% – be better off and 80% in the middle group (FEWS [2004]). This is better than the national pastoral poverty incidence of 28% recorded by Rass ([2006]) for Nigeria (the lowest in West Africa, compared to 48% in neighbouring Chad and Niger, which had the highest poverty incidence amongst pastoralists in West Africa). Despite this comparatively well-to-do status, most of the study households' wealth is invested in their livestock. This means that their day-to-day living standards are lower than would be expected and that their livelihoods are vulnerable to the risks of disease, droughts and conflict.

Crops

The pastoral Fulani group in the study area have become agro-pastoralists who own land and grow their own staples (maize, millet) as well as cash crops (vegetables). They are even willing to invest cash in agricultural inputs such as herbicides and fertiliser. The successful adaptation to agro-pastoralism has resulted in higher food security and income diversification for the Fulani on the Jos Plateau. Typically, pastoralists are highly dependent on cash purchase of food to satisfy their nutritional need, but the widespread adaptation to agro-pastoralism reduces this dependency and frees up cash for other uses. The adaptation to agro-pastoralism and reduced dependence on food purchase is also an indication of wealth status. Poorer pastoralists typically purchase 70% of their food requirements whereas those who are better off only purchase 15% of requirements with 80% being supplied by their own crops (FEWS [2004]).

This transition may be due to the breakdown of viable exchange relations between pastoralists and local farmers, forcing the Fulani to acquire land and grow their own food, instead of getting it from farmers in exchange for dairy products and manure. It may also simply be a natural income diversification strategy enabled by the long-term settlement of the Fulani in this area. This is almost certainly the case in the 23% of households who own large amounts of land and sell most of their crops for cash. However, it is likely that the former reason is responsible for the widespread uptake of arable agriculture observed. It is interesting to note that the Fulani, along with local farmers, mostly grow crops that provide little or no residues (hay and threshed grain heads) for animals to graze.

Herd Composition

Herds were composed of 76% females and 24% males. Further analysis revealed that there were slightly more female calves aged <6 months and equal numbers of juveniles aged 6 months to 2 years, but in adult cattle aged >2 years, the number of females was six times the number of males. This shows that equal numbers of male and female calves are born but that a large number of males exit the herd when they reach the age of two. The high percentage of adult females in the herd is characteristic of a dairy herd, where the husbandry system is geared towards maximising milk production and herd growth. The preponderance of adult females in the herd is a consistent feature of milk- and herd growth-oriented extensive cattle production systems throughout Africa (Pullan [1979]; Wagenaar et al. [1986]; Agyemang et al. [1997]; Roderick et al. [1998]) and contrasts with animal traction-oriented cattle production systems, where numbers are similar or males predominate (Ahmedin and Hugh-Jones [1995]; Ocaido et al. [2005]). However, in our study, the proportion of females in the herd, at 74%, is high as compared to 70% (Roderick et al.[1998]), 70% (Agyemang et al. [1997]) and 67% (Wagenaar et al. [1986]) where some 4% – 6% of the herd were oxen, probably used for draught and, in the same area, 71% (Pullan [1979]). This does indicate that in the study area, herders keep very few surplus males, fewer than in 1979 and fewer than elsewhere in the region. Mature males are preferentially sold when cash is needed.

Herd Productivity

Reproductive efficiency was low but not untypical for the region. Cows in the region usually have their first calf at age 4 years and over - with Pullan ([1979]) finding a value of 60 months, Agyemang et al. ([1997]) 50 to 53 months and Wagenaar et al. ([1986]) 50 months. Thus, it is estimated that between 38% and 45% of the observed population in our study consisted of breeding females. This proportion is very similar to that for other African pastoral systems (Wagenaar et al. [1986]; Agyemang et al. [1997]; Roderick et al. [1998]). The annual calving rate implied by the herd composition would thus be between 33% and 42%. This is substantially lower than the 46% estimated by Pullan ([1979]). The optimum calving rate expected in intensive commercial farming systems is 100%, i.e. one calf per cow per year. However, results from studies on Fulani herds under similar traditional management lead us to expect that 50% calving rate in these circumstances reflects a number of factors. In these production systems, low calving rates are usually attributed to the practice of allowing calves to suckle their dams for very long periods of up to two years. This is done for two main reasons: first, to provide robust nutrition for calves, given the uneven availability of pasture, and second, to ensure a constant supply of milk for sale and household needs. It has also been argued that calving every other year is sustainable for the cows in extensive systems, with limited grazing and some transhumance (Hill [1995]). Lactation length is a significant factor in the determination of calving interval because the stress of lactation delays the onset of oestrus until the calf is weaned. However, here the impact is particularly marked. The effects of disease and poor nutrition exacerbate the situation, so that this means that on average, each cow produces a calf once every 28 to 36 months. The average calving interval for White Fulani cows in Nigeria is 13 to 14 months under intensive management (Wheat and Broadhurst [1968]; Wheat et al. [1972]; Oyedipe et al. [1982]) and 24 to 27 months under traditional management (Pullan [1979]; Otchere [1983]), as against 20 months in Mali's Niger Delta (Wagenaar et al. [1986]) and 22 months in central Mali (Wilson [1986]) as well as The Gambia (Agyemang et al. [1997]) where it was observed that a subset of cows calved regularly and others very infrequently. Calf mortality in the first six months of life is low at 12.7% compared to expected values of 20% – 40% for the first year (Pullan and Grindle [1980]; Wagenaar et al.[1986]), but this is not enough to compensate for the poor reproductive performance.

Mortality was low at 3.5%, much lower than the 16.3% recorded by Wagenaar et al. ([1986]) and 8.1% recorded by Pullan and Grindle ([1980]). This may be because of generally lower levels of disease than in the past as well as increased access of herders to curative drugs.

Sales were equivalent to 3.2% of the herd, again lower than the 5.6% recorded by Wagenaar et al. ([1986]) and 10.4% by Pullan and Grindle ([1980]). This may be because the Fulani on the Jos Plateau are now all agro-pastoralists and are less dependent on cash from sale of cattle to buy food. However, herds are much larger now, and more money is realised per sale, even after allowing for inflation, so selling a smaller percentage of the overall herd is sufficient. Purchases of new stock were equivalent to 1.0% of the herd, comparable to the 1.2% recorded by Pullan and Grindle ([1980]).

Natural herd growth (often described as potential offtake) was 1.6% and actual offtake 3.6%. Both figures are very low, much lower than the 14.7% natural herd growth and 11.7% actual offtake previously recorded by Pullan ([1980]). This points to reduced productivity over time. The reduction in productivity is probably linked to increased land pressure and environmental degradation and its knock-on effects on cattle nutrition and management. It may also be linked to the presence of trypanosomiasis which was absent from the Plateau at the time of the study by Pullan (Majekodunmi et al. [2013a]) which is primarily a disease of production. Actual offtake was more than potential offtake, indicating that herders are unable to meet their need for cash from livestock sales without reducing the size of their herd. This factor is probably responsible for closing stock levels (average herd size of 249) being lower than starting stock (253) and the −1.2% average change in herd size.

Figures obtained previously from Pullan and Grindle ([1980]) were compared to those collected in this survey. There were no statistically significant differences between the two data sets. However, there are some apparent changes that can be deduced from the figures:

Mean herd size has increased and is more variable. This could be due to increased polarisation into better off versus poor households or to a simple increase in the number of better off households. Wealth grouping results support the latter because although there is a higher proportion of better off households, there is also a high proportion of households in the middle category.

Variability in annual change in herd size has decreased, showing that herd sizes are more stable.

The number of births is less variable, but seems to be decreasing, pointing to a reduction in cattle productivity.

Variability in both natural herd growth (potential offtake) and actual offtake has decreased, pointing to increased stability. However, actual offtake in this study is more than potential offtake, which also indicates reduced productivity and ultimately a willingness to reduce herd size and compromise long-term herd growth, if viable breeding females are sold.

Sale and purchase prices are both less variable. The price of stock bought is much lower than it was, but on average, herders earn USD 120 more per animal sold.

When taken together, these factors indicate that the overall livestock enterprise is more stable but the overall productivity has also decreased.

Effects of Conflict, Competition and Migration

Results show that 67% of all households practised transhumance and that over half of these practised both wet and dry season migration. Poor availability of resources is the main driver for dry season migration, being the reason cited by 71% of respondents, almost twice the number who cited poor access as the reason for migration. However, poor access to available resources is the major driver for wet season migration, being the reason cited by 48% of respondents, almost four times the number (13%) who cited lack of resources as the reason for migration.

It is easy to see how the poor pasture and dwindling water supplies caused by the long harsh dry season force herders to move their cattle to areas where the dry season is shorter and its effects less severe. This is an age-old feature of pastoral life. Wet season migration is an unfortunate consequence of natural resource competition and marginalisation of pastoralists by farming communities. The issues surrounding lack of access to abundant pasture and water during the wet season are more involved and require a study of the relationships between Fulani pastoralists and the indigenous farming peoples.

The most significant feature of rural dynamics in Nigeria in the twentieth century has been demographic expansion and the consequent expansion of cultivation (Blench [2003]). Agricultural expansion on the Jos Plateau has also been fuelled by technological advances: improved seed varieties, irrigation pumps and the introduction of new crops like maize and potatoes and vegetable gardening (Blench [2003]). Farmers on the Plateau previously practised rainfed agriculture based on cereals that grew well on eroded soils and produced high-quality residues for animals. This facilitated a stable system of exchange between farmers and pastoralists: grain for dairy products and crop residues for manure. At this time, mine ponds and river valleys were of little use to farmers but were valuable to pastoralists who were given free access but not ownership rights. Dry season farming of new crops - maize, sugarcane and Irish potatoes - was introduced in the 1960s and greatly encouraged by World Bank-funded agricultural development programmes in the 1970s. Year-round large-scale vegetable gardening also began in the 1980s, assisted by the appearance of small petrol irrigation pumps. These are all fertiliser-based crops that produce no useful crop residues for cattle feed, so the farmer-pastoralist exchange system has broken down. Advances in livestock management and disease control over time have also led to agricultural expansion on the part of pastoralists. This has led to overstocking and overexploitation of pasture. Very large numbers of cattle have caused degradation on a massive scale, and sheet and gully erosion have long been widespread on the Jos Plateau (Pullan [1980]; Blench [2004]).

Given the increasing land pressure and agricultural expansion by both parties, previous land use schemes became unsustainable and have broken down. All riverine plots which previously provided lush pasture and watering points have become extremely valuable to farmers. Therefore, pastoralists' access rights have been revoked and they have been effectively excluded from the enjoyment of natural resources that were previously held in common. Pastoralists have adapted to this increasing marginalisation by sending their cows off the Plateau for longer periods each year. There are several important consequences of the increase in transhumance in response to these issues. Management becomes more difficult as herds are split and increased labour is required to herd migrating cattle (which are also more vulnerable to security risks). Cattle suffer poorer health and productivity due to malnutrition, stress and fatigue of trekking, increased morbidity of underlying diseases and increased contact with disease and disease vectors during transhumance. There are also many long-term, indirect effects of this climate of recurring violence, marginalisation of pastoralists and constant risk of conflict. These factors often reverse ongoing development processes in pastoral livelihoods - herders are less likely to invest in new assets/technologies or to practise income diversifications in such an atmosphere of uncertainty (Mahama et al. [2005]; Majekodunmi et al. [2013b]). The outbreaks of urban violence in Jos and its environs exacerbate the growing natural resource conflict in the rural areas.

Environmental security scholars believe that these rural conflicts are manifestations of herder-farmer conflict based on competition for land and natural resources, fuelled by climate change such as desertification in the Sahel (Homer-Dixon [1999]; Herskovits [2010]). Political ecologists on the other hand argue that social, cultural and political considerations are more important than ecological factors in causing these violent conflicts (Hartmann [2001]; Moritz [2010]).

Our experience has been that natural resource conflicts unrelated to the ethno-religious conflicts previously described tend to be settled amicably on the Jos Plateau as reported by respondents. Traditional conflict mitigation/resolution mechanisms are active there and have successfully managed to develop schemes for the shared use of key resources. The pastoralists interviewed described practising increased migration during both wet and dry seasons to comply with these agreements. This is reinforced by Blench ([2004]) who points out that natural resource conflicts on the Jos Plateau are 'usually settled informally and the types of violent clashes characteristic of some other northern states were not characteristic of the Plateau'. He presents resource use maps as evidence of agreements on natural resource use between farmers and herders which were still in use at the time of this study (Figure 13). Higazi ([2013]) is of the opinion that 'the complexity of the rural violence has tended to be lost on analysts and reporters, who in seeking to explain what is happening from afar have over-simplified the conflicts as "farmer-herder clashes" - there are other important aspects to the violence'. These include revenge for past attacks, economies of conflict (cattle rustling and armed robbery for gain), resource scarcity, decreasing interdependence of pastoral and agricultural economies, cultural/religious differences between herders and farmers and institutional failures to resolve conflicts as well as the broader political and historical context (Blench [1984]; Huraoult [1998]; Moritz [2006]; Tonah [2000]; Blench[2004]; Higazi [2011]).

Figure 13: Gashish wet season land use scheme (Google Earth, 2009).

CONCLUSION

This survey of pastoral livelihoods of Fulani herders is set against a backdrop of agricultural change and general insecurity. The survey showed that the majority of pastoral households in the study area are better off or in the middle wealth categories and only a few are poor in terms of livestock holdings. However, many households are poor in terms of land holdings. Income diversity is significant but as many as 30% of households still rely on one source of cash revenue: livestock sales. Pastoralists are less vulnerable to livestock disease but increasingly vulnerable to the risks of conflict and poor access to natural resources. Cattle productivity parameters are more stable, but the overall productivity has decreased and reproductive rates were particularly low. The amount of wealth concentrated in livestock, number of land-poor households and those still reliant on one source of income suggest that there is still a degree of vulnerability to the risks of disease, drought/dry season and conflict in pastoral livelihoods in the area.

This information gives us an improved understanding of factors affecting pastoral livelihood strategies and lifestyle choices, which is useful for targeting improvements and interventions and providing a solution to the current volatile situation. The target is clearly not poverty or livestock disease, although there is room for improvement on both. The target should be to increase productivity and sustainability and reduce vulnerability. The primary approach to achieving this is to settle the issues surrounding security, natural resource use and sustainable nutrition for cattle. The current extensive production system has clearly become unsustainable, and the extent to which Fulani pastoralists can adapt their livelihood strategies to integrate with prevailing conditions will determine the viability of their livelihoods in the future.

AUTHORS' CONTRIBUTIONS

AOM conceived of the study and participated in its design, coordinated the fieldwork, analysed the data and drafted the manuscript. AF and CD carried out the data collection and translations in the field. APMS and SCW participated in the study design and coordination, data analysis and manuscript revision. All authors read and approved the final manuscript.

ACKNOWLEDGEMENTS

We would like to thank Augustine Igweh, Everestus Yanan, Musa Sule and the Nigerian Institute for Trypanosomiasis in Vom, Nigeria, and The Plateau State Ministry of Agriculture for their assistance facilitating this project. We would also like to thank Dr K. Picozzi and Dr M. Thrusfield for their contributions to the wider project. Our sincere appreciation goes to the community leaders of the study villages on the Jos Plateau.

REFERENCES

1. Agyemang, K, RH Dwinger, DA Little, and GJ Rowlands. 1997. Village N'Dama cattle production in West Africa: Six years of research in The Gambia. International Livestock Research Institute/Kenya and International Trypanotolerance Centre, Nairobi/Banjul.
2. Ahmedin, J, and ME Hugh-Jones. 1995. Association of tsetse control with health and productivity of cattle in the Didessa Valley, western Ethiopia. Preventive Veterinary Medicine 22: 29–40.
3. Awogbade, MO. 1983. Fulani pastoralism: Jos case study. ABU Press, Zaria.
4. Bawa, I, and V Nwogwu. 2002. The Jos crisis. In Hope betrayed? A report on impunity and state-sponsored violence in Nigeria, ed. Sottas E, Chukwuma I, Odinkalu C, World Organisation Against Torture/Centre for Law Enforcement Education (CLEEN), Geneva/Lagos.
5. Blench, R. 1984. Conflict and cooperation: Fulani relations with the Samba and Mambila peoples. Cambridge Anthropology 9 (2): 42–57.
6. [www.rogerblench.info] Blench, R. 2003. The transformation of conflict between pastoralists and cultivators in Nigeria. . Accessed 20 June 2009.

7. Blench, R. 2004. Natural resource conflicts in North-Central Nigeria: A handbook and case studies. Mallam Dendo Ltd., Cambridge.
8. Catley, A. 2000. Notes on pastoral livelihoods in Africa. RWA International/Vetwork UK, London.
9. 2007. Economic and financial review. Federal Government of Nigeria, Abuja.
10. Chambers, R. 1994. The origins and practice of participatory rural appraisal. World Development 22 (7): 953–969.
11. [www.eldis.org/go/livelihoods] Eldis.org. 2010. Livelihoods connect. . Accessed 1 May 2011.
12. [http://www.fews.net/sectors/medios-de-vida] FEWS NET. 2004/5. Famine early warning system: National livelihood profiles. . Accessed 19 June 2009.
13. Fricke, W. 1979. Cattle husbandry in Nigeria: A study of its ecological conditions and social-geographical differentiation. Heidelberger Geographischen Arbeiten, Heft 52. Geographisches Institut der Universität, Heidelberg.
14. Gunn, HD. 1953. Peoples of the plateau area of Northern Nigeria. Ethnographic survey of Africa, Western Africa, 7. International African Institute, London.
15. Hartmann, B. 2001. Will the circle be unbroken? A critique of the Project on Environment, Population, and Security. In Violent environments, ed. Peluso NL, Watts M, 39–62. Cornell University Press, Ithaca.
16. [http://allafrica.com/stories/201003150001.html] Herskovits, J. 2010. Nigeria: Critical time for Nigeria's future. Guest Column, 14 March 2010. . Accessed 15 Sept 2014.
17. Higazi, A. 2008. Social mobilization and collective violence: Vigilantes and militias in the lowlands of Plateau State, Central Nigeria. Africa: Journal of the International African Institute 78 (1): 107–135.
18. Higazi, A. 2011. The Jos crisis: A recurrent Nigerian tragedy. Discussion Paper No. 2. Friedrich Ebert Stiftung, Abuja.
19. Higazi, A. 2013. Rural insecurity on the Jos Plateau, Nigeria: Livelihoods, land, and religious reform among the Berom, Fulani, and Hausa. Nigeria Research Network, Oxford.

20. Hill, DH. 1995. The cattle and buffalo of Africa: Management. Outlook on Agriculture 24 (3): 155–161.
21. Homer-Dixon, TF. 1999. Environment, scarcity, and violence. Princeton University Press, Princeton.
22. 2001. Nigeria: Jos, a city torn apart Human Rights Watch 13 (9A): 1–25.
23. Huraoult, J. 1998. Land crisis on the Mambila Plateau of Nigeria, West Africa. Journal of Biogeography 25 (2): 285–299.
24. Krause, J. 2011. A deadly cycle: Ethno-religious conflict in Jos, Nigeria. Small Arms Survey, Geneva.
25. Little, PD. 1989. The dairy commodity system of the Kismayo region, Somalia: Rural and urban dimensions. Institute for Development Anthropology, Binghamton.
26. Mahama, CI, M Desquesnes, ML Dia, B Losson, R De Deken, N Speybroeck, and S Geerts. 2005. A longitudinal epidemiological survey of bovine trypanosomosis and its vectors in the White Volta river basin of Northern Ghana. Veterinary Parasitology 128: 201–208.
27. Majekodunmi, AO, A Fajinmi, C Dongkum, K Picozzi, MV Thrusfield, and SC Welburn. 2013. A longitudinal survey of trypanosomiasis on the previously tsetse-free Jos Plateau, Nigeria: Prevalence, distribution and risk factors. Parasites & Vectors 6: 239.
28. Majekodunmi, AO, A Fajinmi, C Dongkum, K Picozzi, E MacLeod, MV Thrusfield, AP Shaw, and SC Welburn. 2013. Social factors affecting seasonal variation in bovine trypanosomiasis on the Jos Plateau, Nigeria. Parasites & Vectors 6: 293.
29. Michael, BJ. 1987. Milk production and sales by the Hawazma (Baggara) of Sudan: Implications for gender roles. Research in Social Anthropology 9: 105–141.
30. Moritz, M. 2006. Changing contexts and dynamics of farmer-herder conflicts across West Africa. Canadian Journal of African Studies 40 (1): 1–40.
31. Moritz, M. 2010. Understanding herder-farmer conflicts in West Africa: Outline of an analytical approach. Human Organization 69 (2): 138–148.

32. Niamir, M. 1982. Report on animal husbandry among the Ngok Dinka of the Sudan. Rural Development Studies, Harvard Institute for International Development, Cambridge.
33. Ocaido, M, CP Otim, NM Okuna, J Erume, C Ssekitto, RZO Wafula, D Kakaire, J Walubengo, and J Monrad. 2005. Socio-economic and livestock disease survey of agro-pastoral communities in Serere County. Uganda. Livestock Research for Rural Development Soroti District 17: 1–15.
34. Otchere, E. 1983. The productivity of White Fulani (Bunaji) cattle in pastoralist herds in the Kaduna Plains of Nigeria. International Livestock Centre for Africa, Nigeria.
35. Oyedipe, EO, V Buvanendran, and LO Eduvie. 1982. Some factors affecting the reproductive performance of White Fulani (Bunaji) cattle. Tropical Agriculture 59: 231–234.
36. Pullan, NB. 1979. Productivity of White Fulani cattle on the Jos Plateau, Nigeria. I: Herd structures and reproductive performance. Tropical Animal Health and Production 11: 231–238.
37. Pullan, NB. 1980. Productivity of White Fulani cattle on the Jos Plateau, Nigeria. II: Nutritional factors. Tropical Animal Health and Production 12: 17–24.
38. Pullan, NB, and RJ Grindle. 1980. Productivity of White Fulani cattle on the Jos Plateau, Nigeria. IV: Economic factors. Tropical Animal Health and Production 12: 161–170.
39. Rass, N. 2006. Policies and strategies to address the vulnerability of pastoralists in sub-Saharan Africa. Pro-Poor Livestock Policy Initiative, Food and Agriculture Organisation of the United Nations, Rome.
40. Roderick, S, P Stevenson, and J Ndungu. 1998. The production parameters influencing the composition and structure of pastoral cattle herds in a semi-arid area of Kenya. Animal Science 66: 585–594.
41. Tonah, S. 2000. State policies, local prejudices and the cattle rustling along the Ghana - Burkina Faso border. Africa 70 (4): 551–567.
42. Wagenaar, KT, A Diallo, and AR Sayers. 1986. Productivity of transhumant Fulani cattle in the inner Niger Delta of Mali. ILCA

Research Report No. 13. International Livestock Centre for Africa, Addis Ababa.

43. Wheat, JD, and J Broadhurst. 1968. An analysis of data on Bunaji Cattle at Birnin Kudu and Kabomo, N. Nigeria. Samaru Miscellaneous Paper No. 25. Ahmadu Bello University, Zaria.

44. Wheat, JD, PN de Leeuw, and BA Koch. 1972. Bunaji cattle at the Shika Research Station, North Central State, Nigeria. Ahmadu Bello University, Zaria.

45. Wilson, RT. 1986. Livestock production in central Mali: Long-term studies on cattle and small ruminants in the agropastoral system. ILCA Research Report No. 14. International Livestock Centre for Africa, Addis Ababa.

Chapter 3

Reducing Risk from Lahar Hazards: Concepts, Case Studies, and Roles for Scientists

Thomas C Pierson[1], Nathan J Wood[2], and Carolyn L Driedger[1]

[1]U.S. Geological Survey, Volcano Science Center, Cascades Volcano Observatory, 1300 S.E. Cardinal Court, Suite 100, Vancouver 98683, WA, USA

[2]U.S. Geological Survey, Western Geographic Science Center, 2130 SW 5th Ave., Portland 97201, OR, USA

ABSTRACT

Lahars are rapid flows of mud–rock slurries that can occur without warning and catastrophically impact areas more than 100 km downstream of source volcanoes. Strategies to mitigate the potential for damage or loss from lahars fall into four basic categories: (1) avoidance of lahar hazards through land-use planning; (2) modification of lahar hazards through engineered protection structures; (3) lahar warning systems to enable evacuations; and (4) effective response to and recovery from lahars when they do occur. Successful application of any of these strategies requires an accurate understanding and assessment of the hazard, an understanding of the applicability and limitations of the strategy, and thorough planning. The human and institutional components leading to successful application can be even more important: engagement of all stakeholders in hazard education and risk-reduction planning; good communication of hazard and risk information among scientists, emergency managers, elected officials, and the at-risk public during crisis and non-crisis periods; sustained response training; and adequate funding for risk-reduction efforts. This paper reviews a number of methods for lahar-hazard risk reduction, examines the limitations and tradeoffs, and provides real-world examples of their application in the U.S. Pacific Northwest and in other volcanic regions of the world. An overriding theme is that lahar-hazard risk reduction cannot be effectively accomplished without the active, impartial involvement of volcano scientists, who are willing to assume educational, interpretive, and advisory roles to work in partnership with elected officials, emergency managers, and vulnerable communities.

BACKGROUND

Lahars are discrete, rapid, gravity-driven flows of saturated, high-concentration mixtures containing water and solid particles of rock, ice, wood, and other debris that originate from volcanoes (Vallance [2000]). *Primary* lahars are triggered during eruptions by various eruption-related mechanisms; between AD 1600 and 2010 such lahars killed 37,451 people worldwide, including 23,080 in the 1985 Nevado del Ruiz disaster alone (Witham [2005]; Aucker et al. [2013]). During the same period *secondary* lahars, most commonly triggered by post-eruption erosion and entrainment of tephra during heavy rainfall, killed an additional 6,801 (Aucker et al. [2013]). Just in the past several decades, staggering losses from widely publicized lahar-related disasters at Mount St. Helens, USA; Nevado del Ruiz, Colombia; Mount Pinatubo, Philippines; and Mount Ruapehu, New Zealand, have demonstrated how lahars of both types significantly threaten the safety, economic well-being, and resources of communities downstream of volcanoes. Lahars can range in consistency from thick viscous slurries resembling wet concrete (termed *debris flows*) to more fluid slurries of mostly mud and sand that resemble motor oil in consistency (termed*hyperconcentrated flows*). These two types of flows commonly occur in all types of mountainous terrain throughout the world, but the largest and most far-reaching originate from volcanoes, where extraordinarily large volumes of both unstable rock debris and water can be mobilized (Vallance and Scott [1997]; Mothes et al. [1998]).

The destructive nature of lahars derives from their speed, reach, and composition—and our difficulty in predicting (in the absence of warning systems) when they may occur. Large lahars commonly achieve speeds in excess of 20 m/s on the lower flanks of volcanoes and can maintain velocities in excess of 10 m/s for more than 50 km from their source when confined to narrow canyons (Cummans [1981]; Pierson [1985]; Pierson et al. [1990]) (Table 1). Impact forces from multi-ton solid objects commonly suspended in debris-flow lahars (such as large boulders, logs, and other debris) and drag forces exerted by the viscous fluid phase can destroy almost any structure (Figure 1a). Hyperconcentrated-flow lahars damage structures primarily through vigorous lateral erosion of channels that results in bank collapse (Figure 1b). Both flow types commonly occur during a single lahar event as the highly concentrated head of a lahar typically transitions to a more dilute tail. On flow margins or at the downstream ends of depositional zones where velocities are much slower, lahars can encase buildings, roads, towers, and farm land in mud-rock slurries that can dry out to near concrete-like hardness. Yet fresh lahar deposits, commonly many meters deep, can remain fluidized like quicksand for days to weeks, complicating search and rescue efforts. Although most lahars are triggered during or shortly after volcanic eruptions, they can also be initiated without warning by noneruptive events, such as the gravitational collapse of structurally weakened volcanic edifices, large earthquakes, lake outbreaks, or extreme rainfall.

Table 1: Examples of lahar travel times from lahar source areas (points of initiation) to selected locations in downstream river valleys

Lahar date	Specified locations	Lahar trigger	Distance of specified location from lahar source (km)	**Travel time from lahar source to specified location (min)**	**Average speed (meters per second)[1]**
1926	Points along Hurano River, downstream of vent at Tokachidake volcano, Japan	Eruption (pyroclastic density current on snow and ice)	2.4	1	42
			6.4	4	26
			20.0	17	16
			23.2	29	5
1980	Points along Pine Creek, downstream of vent at Mount St. Helens, USA	Eruption (pyroclastic density current on snow and ice)	10.1	7	24
			14.1	11	16
			16.8	13	18
			20.9	19	14
1985	Point along Denjo River, downstream of landslide at Mount Ontake volcano, Japan	Earthquake-triggered slope failure	11.4	10	19
1985	Points along Río Chinchiná, downstream of vent at Nevado del Ruiz volcano, Colombia	Eruption (pyroclastic density current on snow and ice)	11.3	20	10
			33.0	70	7
			68.6	172	6
1990	Points along Drift River, downstream of vent at Redoubt Volcano, Alaska	Eruption (pyroclastic density current on snow and ice)	7.6	8	16
			18.5	27	10
1994	Points along Río Páez, downstream of vent at Nevado del Huila volcano, Colombia	Earthquake-triggered slope failures	4	2 to 3	22 - 33
			9	6 to 9[2]	17 - 25
			30	20 to 30[2]	17 - 25

Travel times are partly a function of lahar magnitude and are determined from eyewitness accounts, instrument recordings, and back-calculations of flow velocity based on physical evidence. Data from Pierson ([1998]) and Scott et al. ([2001]).

[1]Average speed is computed from the closest upstream point where timing information was available; not all timing points shown in this table. Localized lahar velocities tend to decrease as channel slopes decrease with distance downstream.

[2]Lahar could be heard or felt about 5 minutes before its arrival at this point.

Pierson et al.

Pierson et al. Journal of Applied Volcanology 2014 3:16 doi:10.1186/s13617-014-0016-4

Figure 1: Destructive effects of lahars. (a) Aerial view of Armero, Colombia, following destruction by a lahar on November 13, 1985, that killed approximately 21,000 people at this site alone (see Pierson et al.[1990]; USGS photo by R.J. Janda, 9 Dec 1985). Patterns of streets and building foundations are visible in the debris field at center of photo. (b) Aerial view of part of Angeles City, downstream of Mount Pinatubo, Philippines, along the Abacan River, showing consequences of vigorous bank erosion by repeated post-eruption hyperconcentrated-flow lahars that were triggered by heavy monsoon rains (see Major et al. [1996]; USGS photo by TCP, 15 Aug 1991).

Various approaches to reduce and manage societal risks associated with lahar hazards have been applied over the years (Neumann van Padang [1960]; Smart [1981]; Suryo and Clarke [1985]; Pierson [1989]). These approaches fall into four basic categories of mitigation, including hazard avoidance, hazard modification, hazard warning, and hazard response and recovery (Figure 2). The goal of this paper is to provide an overview of each of these risk-reduction strategies and to highlight case studies of how (and how effectively) they have been applied at volcanoes around the world. The timing and magnitude of future lahars is uncertain and risk reduction efforts can be financially and politically costly; therefore economic, political, and social factors can compromise the implementation and long-term effectiveness of any strategy (Voight [1990], [1996]; Newhall and Punongbayan [1996];

Peterson [1996]; Prater and Lindell [2000]). We begin by discussing the importance of hazard and risk education for affected populations, elected officials, and emergency managers. We end by reemphasizing the call for committed involvement by volcano scientists in developing and executing these strategies. Scientist involvement improves the credibility and the efficacy of risk-reduction efforts. When the risks are perceived as credible and risk-reduction strategies are understood, tragic losses from future lahars on the scale of 20th-century lahar disasters can be avoided or at least minimized.

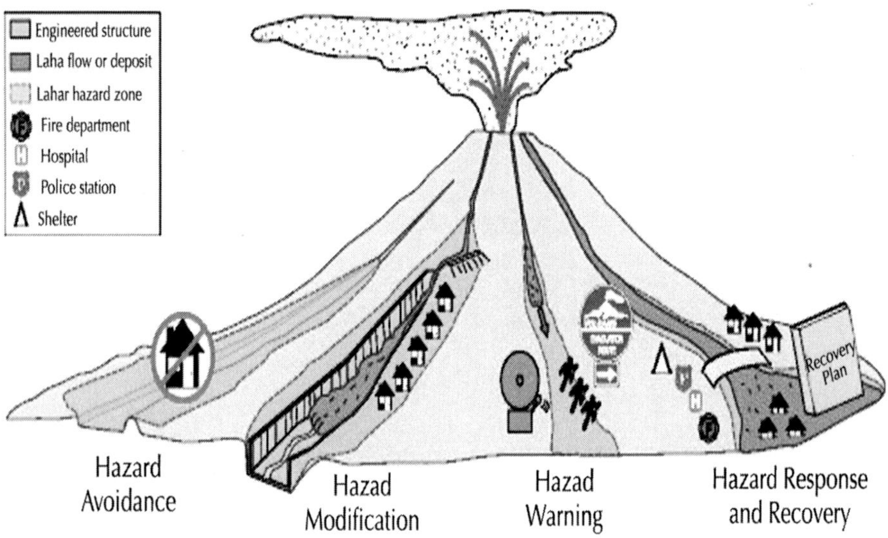

Figure 2: Schematic representation of the four basic strategies to reduce lahar-hazard risk within lahar hazard zones. Strategies include (1) hazard avoidance with land-use planning and zonation; (2) hazard modification with engineered protection structures (bypass channel and deflection berm); (3) hazard warning to allow for timely evacuation; and (4) hazard response and recovery, which minimize long-term impacts after a lahar has occurred.

HAZARD AND RISK EDUCATION

The foundation for all risk-reduction strategies is a public that is well informed about the nature of hazards to their community, informed about how to lessen societal risk related to these hazards, and motivated to take risk-reducing actions. This knowledge base and accompanying appreciation of volcano hazards are needed to increase the interest and ability of public officials to implement risk-reduction measures and create a supportive and responsive at-risk population that will react appropriately when an extreme event occurs. Volcano scientists play a critical role in effective hazard education by informing officials and the public about realistic hazard probabilities and scenarios (including potential magnitude, timing, and impacts); by helping evaluate the effectiveness of proposed risk-reduction strategies; by helping promote acceptance of (and confidence in) hazards information through participatory engagement with officials and vulnerable communities as partners in risk reduction efforts; and by communicating with emergency managers during extreme events (Peterson [1988], [1996]; Cronin et al. [2004b]; McGuire et al.[2009]). But before successful use of hazard information can occur, the scientists' first and main role is to make technical data, hypotheses, and uncertainties understandable to non-technical users of hazard information. Serious misunderstandings can arise, sometimes with tragic consequences, when scientists do not perform this role effectively (Voight [1990]; Hall [1992]).

An effective hazard education program begins when scientists inform people in vulnerable communities about past hazardous events and current threats—information necessary for preparedness for future events. Scientists need to be involved in hazard-education efforts, because they provide the needed hazard expertise, and the public tends to imbue them with a high level of trust (Ronan et al. [2000]; Haynes et al. [2008]; Mei et al. [2013]). But the straightforward presentation of information that may seem logical to many scientists may not be effective; hazards information must be transmitted in ways that are not only understandable but also emotionally palatable and culturally relevant to the target audience (Cronin et al. [2004b]). People are more likely to implement risk-reduction strategies before an event or evacuate during an event if they comprehend that past events have impacted their communities, if they believe that future events could do so again and that viable mitigation options exist, and if they themselves have been involved in determining their community's risk-reduction strategies (Mileti [1999]). Community adoption of mitigation strategies is also more likely if hazard education is integrated into existing development programs and if it includes discussion of tangible actions that can be taken to protect lives and livelihoods, instead of just discussing uncontrollable threats (Paton et al. [2001]). The types of educational products, activities, and tasks that benefit from the active participation of scientists are varied (Figure 3):

Informative, jargon-free, general-interest publications and multi-media information products about potential hazards in digital and print formats (e.g., IAVCEI [1995], [1996]; USGS [1996], [1998],[2010]; Gardner et al. [2000]; Gardner and Guffanti [2006]; Driedger and Scott [2008]; Dzurisin et al. [2013]).

Technical information products to summarize scientific information about potential or ongoing volcanic activity or potential hazards, such as hazard-assessment reports, alerts and information statements on the status of current volcanic activity, volcanic-activity notification services, response plans developed in partnership with other agencies and stakeholders, and specific guidance based on the latest research (Guffanti et al. [2007]). Such products can be made available through print, fax, email, web-site, and social media outlets (e.g., Scott et al. [1997]; Hoblitt et al. [1998]; Pierce County [2008]; Wood and Soulard [2009a]).

Accessible and understandable spatial depictions of hazardous areas and evacuation routes to safe areas that are tailored to a target audience (Figure 3a,b), such as traditional hazard maps, evacuation route maps, explanations of the volcanic origins of familiar landscape features, labeled aerial photographs with vertical and oblique perspectives, and simple perspective maps keyed on cultural features and boundaries (Haynes et al. [2007]; Némath and Cronin [2009]). Web sites developed by local agencies can be good outlets for this type of information (e.g.,http://www.piercecountywa.org/activevolcano *webcite*).

Hazards information presentations and training for the media (Figure 3c), emergency management officials (Figure 3d), first responders, land managers, public safety officials, search-and-rescue (SAR) teams, community-based monitoring teams, and public information officers before and during volcano crises (Driedger et al. [2008]; Frenzen and Matarrese [2008]; Peterson [1988],[1996]; Driedger et al. [2008]; Driedger and Scott [2010]; de Bélizal et al. [2013]; Stone et al.[2014]).

Teacher trainings (Figure 3e) and special school curricula for children in order to provide a foundation of knowledge at a young age, as well as to educate and motivate their families (e.g., Driedger et al. [2014]).

Presentations to and dialogues with community groups and councils, volunteer organizations, local government bodies, and schools about existing hazards (Figure 3f), while seeking opportunities to engage vulnerable populations in devising potential options for risk reduction (Peterson [1988],[1996]; Driedger et al. [1998]; Cronin et al. [2004a],[b]).

Relationship-building with communities and community leaders (official and unofficial) to establish trust and credibility, to encourage community-based risk-reduction solutions, and to maintain an ongoing dialogue with officials and at-risk community members (Peterson [1988], [1996]; Cronin et al. [2004b]; Haynes et al. [2008]; McGuire et al. [2009]; Mileti [1999]; Stone et al. [2014]).

Collaboration with emergency managers in the design and message content of signs for hazard awareness, locations of hazard zones, and evacuation procedures and routes (Figure 3g) (Schelling et al. [2014]; Driedger et al. [1998], [2002], [2010]; Myers and Driedger [2008a], [b]) and for disaster commemorations (such as monuments or memorials) that remind the public that extreme events are possible (Figure 3h).

Collaboration in the development of accurate and consistent warning messages to be sent out when a lahar triggers a warning system alert (Mileti and Sorenson [1990]).

Figure 3: Examples of some approaches for communicating hazards information to emergency managers, public officials, and at-risk populations. (a) *Non-traditional hazard maps*: An oblique perspective map showing potential lahar zones (brown) emanating from Mount Rainier volcano, with City of Tacoma, Washington (79 km downstream of Mount Rainier), in lower center of image along Puget Sound shoreline. Many people find it easier to visualize spatial information on such maps than on vertical plan-view maps. Satellite ground-surface image from Google Earth® modified by NJW, with Case 1 lahar hazard zones from Hoblitt et al. ([1998]) overlaid. (b)*Signs and post-*

ers: A trail sign for hikers, using words and pictures, to convey lahar hazard information and instructions on what to do if they hear an approaching lahar (Mount Rainier National Park, USA). (c)*Working with media*: A USGS-hosted press conference to inform the media about the reawakening of Mount St. Helens (USA) in 2004 (USGS photo by D. Wieprecht). (d)*Training*: A training class on volcano hazards for emergency managers and given by scientists to provide an opportunity for relationship-building, as well as education (USGS photo by CLD). (e)*Working with teachers*: A scientist-led teacher workshop where simple physical models of lahars were used to help teachers grasp (and later teach) fundamental concepts about lahars (USGS photo by CLD). (f) *Involving vulnerable populations in hazard-mitigation decisions*: A 3-dimensional participatory mapping exercise for residents of a threatened village at Merapi volcano, Indonesia (photo by F. Lavigne, used with permission). (g) *Practice drills*: A lahar evacuation drill in 2002 at a school in Orting, Washington, which is downstream of Mount Rainier (USGS photo by CLD).(h) *Monuments and memorials*: A simple disaster memorial commemorating 22 people killed by lahars in the town of Coñaripe on the lower flank of Villarrica volcano, Chile, in 1964 (USGS photo by TCP).

Hazard education materials should be tailored to address the demographics and socioeconomic context of at-risk populations (e.g., Wood and Soulard [2009b]). This may include providing information in multiple languages on signs, pamphlets, and warning messages where appropriate, or conveying information in pictures or cartoons to reach children and nonliterate adults (Ronan and Johnston [2005]; Tobin and Whiteford [2002]; Dominey-Howes and Minos-Minopoulos [2004]; Gavilanes-Ruiz et al. [2009]). Educational outreach should also include efforts to reach tourists and tourism-related businesses, because these groups may lack hazard awareness and knowledge of evacuation procedures (Bird et al. [2010]).

A hazards and risk education program can increase its effectiveness by focusing outreach on those individuals and groups who can further spread information throughout a community. Such outreach can target institutions such as social organizations, service clubs, schools, and businesses, as well as trusted social networks (Paton et al. [2008], Haynes et al. [2008]). The key to sustaining hazard education is to identify and train community members with a vested interest in preparedness, such as emergency managers, educators, health advocates, park rangers, community and business leaders, and interested residents and other stakeholders. Training community members to integrate hazard information into existing social networks is especially crucial for hard-to-reach, potentially marginalized community groups, such as recent immigrants, daily workers coming from outside of hazard zones, or neighborhoods with people who don't speak the primary language (Cronin et al. [2004a]).

Direct involvement in training community members and elected officials extends a scientist's capacity to educate a community. It also provides opportunities for scientists to gain insight on how people conceptualize and perceive the hazards and the associated risks (for example, the role traditional knowledge and local experience), strengths and weaknesses of communication lines within a community, and any context-appropriate measures that might be used to increase local capacity for risk reduction (Cronin et al. [2004b]). Several studies have shown that people's behavior towards volcano risks is influenced not only by hazards information but also by the time since the last hazardous event and the interaction of their perceptions with religious beliefs, cultural biases, and socioeconomic constraints (Lane et al. [2003]; Gregg et al. [2004]; Chester[2005]; Lavigne et al. [2008]). Understanding these influences and the socio-cultural context of risk is important if scientists are to successfully change behaviors and not simply raise hazard awareness. Participatory methods such as three-dimensional mapping (Gaillard and Maceda[2009]) (Figure 3f), scenario planning (Hicks et al. [2014]), participatory rural appraisals (Cronin et al. [2004a][2004b]), and focus group discussions (Chenet et al. [2014]) can be used to understand the societal context of volcanic risk, to integrate local and technical knowledge, and to promote greater accessibility to information. These "bottom-up" efforts, as opposed to government-driven efforts that are perceived as "top-down", promote local ownership of the information (Cronin et al.[2004b]), empower at-risk individuals to implement change in their communities (Cronin et al.[2004a]), and can result in risk-reduction efforts becoming an accepted part of community thinking and daily life.

Finally, scientists should understand that effective hazard and risk education is a long-term investment of time and resources and will not be a one-time effort. One issue is that people may show great enthusiasm in hazards and risk information at public forums, but their interest and participation in risk-reduction activities may diminish over time as other day-to-day issues become higher priorities. Another issue is unavoidable turnover among users of hazards information. Elected officials may retire or be voted out of office. Emergency managers, first responders, and teachers may transfer to other positions or retire. People move in and out of vulnerable communities. So, just as scientists continually monitor changing physical conditions at volcanoes, they should also appreciate the dynamic nature of the perceptions and knowledge of hazards within communities, agencies, and bureaucracies—and plan for sustained education and outreach efforts.

STRATEGIES FOR LAHAR-HAZARD RISK REDUCTION

Each of the four basic risk-reduction strategies of hazard avoidance, hazard modification, hazard warning, and hazard response and recovery (Figure 2) has basic underlying requirements for successful application. These requirements include an accurate assessment of the hazard; a realistic understanding by elected officials, emergency managers, and at-risk populations of the hazards, risks, and limitations of any implemented strategy; thorough planning; adequate funding; practice exercises and drills, where appropriate; and effective communication among stakeholders during actual lahar occurrence (Mileti [1999]; Leonard et al. [2008]). Scientists have important roles to play in all of these underlying requirements.

Hazard Avoidance

A range of approaches can either regulate or encourage hazard avoidance—the strategy seeking to expose as few lives and societal assets as possible to potential loss. Land-use zoning regulations or development of parks and preserves that ban or limit occupation of hazard zones are ways to keep people, developed property, and infrastructure out of harm's way. Another way is for local government policies to allow occupation of hazard zones but to also impose disincentives for those who choose to live there. A third way is to educate the public about the hazard, the risks, and the probabilities of hazardous event occurrence, and then to trust that people will choose to minimize the hazard exposure of their homes and businesses.

A complete ban on development in a hazard zone is probably the most effective way to avoid the hazard. This may be easiest immediately following a disaster and if the ban aligns with cultural values, such as when the entire town site of Armero, Colombia, was made into a cemetary after about 21,000 people were killed there by a lahar in 1985 (Pierson et al. [1990]; Voight [1990]). However, it is commonly challenging to implement development bans based on hazard zonation prior to a disaster due to people's strong attachment to a place, cultural beliefs, political push-back from business and real-estate interests, the lack of alternative locations for new development, attitudes of individuals who don't want to be told where they can or cannot live, or needed access to livelihoods that exist in volcano hazard zones (Prater and Lindell [2000]; Lavigne et al. [2008]). Indeed, lahar hazard zones can be attractive for transportation and other infrastructure and for residential development, because these areas typically encompass deposits of previous lahars that offer flat topography, commonly above flood hazard zones, and they may offer scenic views of a nearby volcano (Figure 4). Lahar and related deposits also may be attractive for resource extraction. In the Gendol valley at Mount Merapi (Indonesia) for example, thousands of people work daily as miners in high-hazard zones, excavating sand and gravel to sell. Most, if not all, are aware of the risk but are willing to accept it because of the financial reward (de Bélizal et al.[2013]). In other cases such hazard zones may already be occupied by well-established communities—a reality that makes development bans problematic. A strong cultural attachment to the land and the lack of available safe land elsewhere may lead communities to accept lahar risks and even continue to rebuild homes after multiple lahar burials (Crittenden [2001]; Crittenden and Rodolfo [2002]).

Figure 4: Mount Rainier volcano and dense residential housing in downstream community of Orting, Washington. The town is built on the flat upper surface of a lahar deposit from Mount Rainier that was emplaced about 500 years ago. Orting is one of several communities that are in lahar hazard zones downstream of Mount Rainier. A warning system in this valley would give residents about 40 minutes to evacuate to high ground (USGS [2013]). USGS photograph by E. Ruttledge, 18 Jan 2014.

A more realistic land-use planning approach may be to restrict the kind or amount of development allowed to occur in lahar hazard zones. For example, vulnerable valley floors could be limited to agricultural use only, with homes built on higher ground. Downstream of Mount Rainier in Pierce County (Washington, USA), comprehensive land use plans include urban growth boundaries that prohibit tourist facilities larger than a certain size and limit other high-density land uses in lahar hazard zones (Pierce County [2014]). Downstream of Soufriére Hills volcano in Montserrat (British West Indies), only daylight entry into certain hazard zones for farming was allowed in the 1990s, due to pyroclastic-flow and lahar hazards associated with the actively erupting volcano (Loughlin et al. [2002]). The goal of such restrictions is to minimize population exposure and to only allow land uses in which people could be evacuated quickly, yet such measures are not always foolproof (Loughlin et al. [2002]). Ordinances can also limit the placement of critical facilities (hospitals, police stations, schools, and fire stations) in hazard zones, so that basic community services would be available for rescue, relief, sheltering, and recovery efforts in the event of a lahar (Pierce County [2014]).

Where no restrictions are imposed on development of lahar hazard zones, it may be possible to discourage development through the use of various disincentives. These could include higher property tax rates, higher insurance rates, and limitation of public services or infrastructure in designated hazard zones. For example in the United States, the National Flood Insurance Program requires that people living in designated flood zones purchase flood insurance (Michel-Kerjan[2010]). As premiums for such types of insurance increase, purchase of a home in a hazard zone should become less attractive.

Hazard education alone could, theoretically, also achieve some hazard avoidance, but evidence suggests that many residents already living in hazard-prone areas rarely undertake voluntary loss-prevention measures to protect their property, despite increased hazard awareness (Michel-Kerjan [2010]). Discouraging new residents from moving into hazard zones may be more realistic. Focused public education campaigns are one way to raise hazard awareness. Another is to require that hazard information be disclosed to people buying property or building structures in a hazard zone. Such disclosures are required on building-permit applications in Orting, Washington in the lahar hazard zone downstream of Mount Rainier. Some individuals may use increased hazard awareness to assess whether the risk is acceptable, others may not, and still other may object to increased hazard awareness. In fact, just the dissemination of hazards information to people living in hazard zones can engender fierce political opposition, particularly from some business and real-estate interests (Prater and Lindell [2000]).

Volcano scientists play important supporting roles throughout any land-use planning process aimed at reducing risk from lahar hazards. First, land-use decisions require hazard-zonation maps that are scientifically defensible, accurate, and understandable, given the potential for political, social, or legal push-back from various constituents. Second, good planning needs input from predictive models that estimate lahar runout distances, inundation areas, and travel times to populated areas. In addition, scientists are needed to help explain the uncertainties inherent in the maps and models, to estimate the likelihood of occurrence, and to evaluate the effectiveness of proposed risk-reduction strategies as land-use planners balance public safety against economic pressures to develop.

Hazard Modification

Some communities predate recognition that they are situated in a lahar hazard zone. Others may expand or be developed in hazard zones because of social and economic pressures, inadequate understanding of the risks, or acceptance and tolerance of the risks. When societal assets are already in lahar hazard zones, construction of engineered protection structures can reduce risk by (a) preventing some lahars from occurring, (b) weakening the force or reach of lahars, (c) blocking or trapping lahars before they can reach critical areas, or (d) diverting lahars away from critical areas—all methods of hazard modification (Smart [1981]; Baldwin et al. [1987]; Hungr et al. [1987]; Chanson [2004]; Huebl and Fiebiger [2005]). Engineered protection works, sometimes referred to as sabo works (*sabō*="sand protection" in Japanese), and slope stabilization engineering methods have been widely used for centuries in volcanic areas in Japan and Indonesia, as well as in the Alps in Europe for protection from nonvolcanic debris flows.

Engineered structures designed for lahar protection downstream of volcanoes have many of the same advantages and disadvantages of river levees in flood-prone areas, sea walls in coastal areas, or engineered retrofits to buildings and bridges in seismic areas. The main advantages of this approach are that communities can survive small- to moderate-size events with little economic impact, and communities, if they choose to, can gradually relocate assets out of hazard zones. However, protection structures are expensive to build and maintain, which may overly burden communities financially or lead to increased vulnerability if funding priorities shift and maintenance is neglected. Another important disadvantage is that protection structures tend to lull populations into a false sense of security. People commonly assume that all risk has been eliminated, and this perception may result in fewer individuals taking precautionary steps to prepare for future events. This view may also result in increased development of areas now perceived to be safe because of the protective structure. The reality is that risk is eliminated or reduced only for events smaller than the 'design event' that served as the basis for construction. Events larger than the design event can occur and when they do, losses can be even larger because of the increased development that occurred after construction of the protection structure—also referred to as the 'levee effect' in floodplain management (Tobin [1995]; Pielke [1999]). This was the case near Mayon Volcano (Philippines) where lahar dikes built in the 1980s led to increased development behind the structures. When they failed because of overtopping by lahars during Typhoon Reming in 2006, approximately 1,266 people were killed (Paguican et al. [2009]). The effectiveness and integrity of engineered structures can also be compromised by the selection of cheap but inappropriate construction materials (Paguican et al. [2009]) and by ill-informed human activities, such as illegal sand mining at the foot of structures or dikes occasionally being opened to allow for easier road access into communities. Therefore, although protection structures may reduce the number of damaging events, losses may be greater for the less frequent events that overwhelm the structures. In addition, engineered channels and some other structures can have negative ecological effects on watersheds.

The potential for large losses is exacerbated if public officials choose to build the structure that is affordable, rather than the structure a community may need. Economics and politics may play a bigger role than science in deciding the type, size, and location of protection structures, because of the high financial costs and land-use decisions associated with building the structures and with relocating populations that occupy construction areas (Tayag and Punongbayan [1994]; Rodolfo[1995]) (Case study 1). Because decision makers will have to balance risk against cost, scientists have a significant role in helping public officials by (a) estimating the maximum probable lahar (the design event); (b) predicting probable flow routes, inundation areas, and possible composition and flow-velocity ranges; (c) estimating probabilities of occurrence; and (d) evaluating the effectiveness of proposed mitigation plans and structures.

Case Study 1. When Economics and Politics Trump Science

Following the June 15, 1991, eruption of Mount Pinatubo (Philippines), lahars and volcanic fluvial sedimentation threatened many downstream communities. Geologists from a number of institutions met with officials at local, provincial, and national levels to explain the threats and to evaluate and discuss proposed countermeasures. Due to political pressures (Rodolfo[1995]*), officials ultimately adopted a lahar mitigation strategy that was based on the construction of parallel containment dikes close to the existing river channels, using easily erodible fresh sand and gravel deposits of earlier lahars as the construction material. Appropriation of the private land needed for lahar containment areas of adequate size was viewed by officials as too politically costly. Officials hoped the dikes would divert lahars and floods past vulnerable communities. However, nearly all the geologists involved in the discussions expressed the opinion that this was a poor strategy because (a) channel gradients were too low for effective sediment conveyance and deposition would occur in the wrong places, (b) dike placement did not provide adequate storage capacity and dikes would be overtopped or breached, (c) most of the dikes were not revetted and would be easily eroded by future lahars, and (d) people would be lured back to live in still-dangerous hazard zones. The advice of the scientists was not heeded, and over the next several years many of these predictions came true, including breached dikes due to lahar erosion and overtopped dikes due to sediment infill. Lahars breaking through the levees caused fatalities and destroyed many homes. A government official later explained (to TCP) that political considerations prompted the decisions to minimize the area of condemned land and build lahar catch basins that were too small. He felt that the plan recommended by the geologists would have angered too many people and that it was better for officials to be seen doing something rather than nothing, even if the chance of success was low. Indeed, political and economic forces can override scientific recommendations (Tayag and Punongbayan*[1994]*; Rodolfo*[1995]*; Janda et al.*[1996]*; Newhall and Punongbayan*[1996]*; Crittenden*[2001]*).*

Slope Stabilization and Erosion Control

Volcanic ash mantling hillslopes is extremely vulnerable to rapid surface erosion and shallow landsliding, and it is easily mobilized as lahars by heavy rain (e.g., Collins and Dunne [1986]; Pierson et al. [2013]). Even after long periods of consolidation and revegetation, ash-covered slopes can fail on massive scales and result in catastrophic lahars (Scott et al. [2001]; Guadagno and Revellino [2005]). Various methods of slope stabilization, slope protection, and erosion control can limit shallow landsliding or surface erosion in disturbed landscapes that could produce extreme sediment inputs to rivers (Figure 5), although most of these approaches are intensive, costly, and generally limited to hillside-scale problem areas (see overviews in Theissen [1992]; Morgan and Rickson [1995]; Gray and Sotir [1996]; Holtz and Schuster [1996]; Schiechtl and Stern [1996]; Beyers [2004]; Valentin et al. [2005]). These are only briefly summarized here. Options for drainage-basin-scale slope stabilization and erosion control are more limited, have been tested mostly in basins disturbed by wildfire rather than by volcanic eruptions, and are not always effective (Beyers [2004]; deWolfe et al. [2008]).

Figure 5: Example of slope stabilization. Timber retaining walls used to stabilize a steep slope in a volcanic area in Japan (USGS photo by TCP).

Regardless of scale of application, slope stabilization and erosion control techniques attempt to either (a) prevent shallow landsliding by mechanically increasing the internal or external forces resisting downslope movement, decreasing the forces tending to drive downslope movement, or both; or (b) prevent rapid surface erosion and sediment mobilization on slope surfaces and in rills, gullies, and stream channels (Gray and Sotir [1996]; Holtz and Schuster [1996]). Inert materials used to stabilize slopes and control erosion include steel, reinforced concrete (pre-cast elements or poured-in-place), masonry, rock, synthetic polymers, and wood, although many of these degrade and weaken with time. Biotechnical stabilization (Morgan and Rickson [1995]; Gray and Sotir[1996]) uses live vegetation to enhance and extend the effectiveness of many engineered structures.

Forces resisting slope failure or erosion can be maintained or augmented by a variety of approaches (Morgan and Rickson [1995]; Gray and Sotir [1996]; Holtz and Schuster [1996]). Counterweight fills, toe berms, retaining walls, and reinforced earth structures can buttress toes of slopes. To maintain buttressing at a toe slope, revetments using riprap, gabion mattresses, concrete facings, and articulated block systems can prevent toe-slope erosion. Anchors, geogrids (typically wire-mesh mats buried at vertical intervals in a slope face), cellular confinement systems consisting of backfilled three-dimensional structural frameworks; micro-piles, deeply rooted woody vegetation, chemical soil binders, and drains to decrease internal pore pressures can increase the shear strength of natural or artificial slopes. To reduce the driving forces, proven methods include regrading to lower slope angles, and weight reduction of structures or materials placed on slopes. Surface erosion of slopes can be controlled by protecting bare soil surfaces and by slowing or diverting surface runoff through the application of reinforced turf mats, geotextile and mulch blankets, hydro-seeded grass cover, and surface drains. Channelized surface erosion can be retarded with gully fills or plugs of cut brush or rock debris, or small check dams.

Intensive slope-stabilization and erosion-control techniques such as many of those listed above may be too costly for large areas of volcanically disturbed drainage basins, but they may be cost-effective in specific problem areas. Over large areas, economically feasible approaches may include tree planting, grass seeding, and grazing management to limit further destruction of slope-stabilizing vegetation.

However, much post-disturbance erosion is likely to occur before grass seed can germinate or tree seedlings can grow to effective size, and a number of studies have shown that large-scale aerial grass seeding is no more effective for erosion control than the regrowth of natural vegetation (deWolfe et al. [2008]).

Lake Stabilization or Drainage

Stabilizing or draining lakes that could breach catastrophically without warning is another way to prevent lahars from reaching vulnerable downstream areas. Crater lakes, debris-dammed lakes (dammed by pyroclastic-flow, debris-avalanche, or lahar deposits), and glacial moraine-dammed lakes all can become unstable if their impounding natural dams are overtopped or structurally fail. Historic rapid lake outbreaks in several countries have triggered catastrophic lahars that resulted in loss of life (O'Shea [1954]; Neumann van Padang [1960]; Umbal and Rodolfo [1996]; Manville[2004]). Very large prehistoric outbreaks of a volcanically dammed lake have been documented having peak flows comparable to the world's largest floods (Scott [1988]; Manville et al. [1999]). Stabilization methods include armoring of existing spillways on natural dams, construction of engineered spillways, and rerouting lake outflow by pumping or drainage through tunnels (Sager and Chambers [1986]; Willingham [2005]) (Figure 6; Case study 2). Preemptive drainage of dangerous lakes can be fraught with difficulties and may not be successful (Lagmay et al. [2007]).

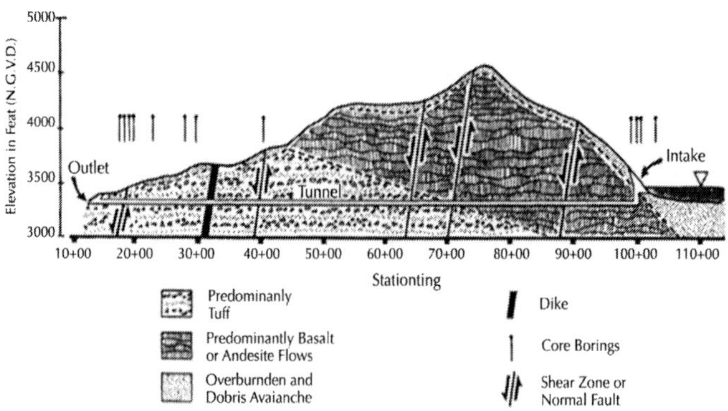

Figure 6: Lake-level stabilization to prevent failure of a natural debris dam and a subsequent lahar. At Mount St. Helens (USA) a tunnel was bored through

a mountain ridge to divert water from Spirit Lake into an adjacent drainage basin. In this case debris-avalanche and pyroclastic-flow deposits formed the potentially unstable natural dam. This geologic cross section shows the 2.5-km-long outlet tunnel, which stabilizes the lake by keeping the water surface at a safe level below the dam crest (from Sager and Budai [1989]).

Case Study 2. Examples of Lake Stabilization

Since AD 1000, 27 eruptions of Mount Kelud (Java, Indonesia) have catastrophically expelled lake water from the volcano's crater lake and created several deadly lahars, including a lahar in 1919 that killed more than 5000 people (Neumann van Padang[1960]*). In an attempt to drain this lake, engineers in 1920 dug a drain tunnel over 955 m in length from the outer flank of the cone into the crater but eventually abandoned the project because of ongoing volcanic activity and other technical difficulties. Thereafter, siphons were constructed to control the lake level, and these were responsible for partial drainage of the crater lake and for a reduced number of lahars during the 1951 eruption (Neumann van Padang*[1960]*).*

More recently, debris-avalanche and pyroclastic-flow deposits from the 1980 eruption of Mount St. Helens (Washington, USA) blocked tributary drainages of the North Fork Toutle River and enlarged several preexisting lakes. The largest and potentially most dangerous of these was Spirit Lake, which, when mitigation efforts began, was impounding 339 million m3of water—enough to form a lahar that could have destroyed major parts of several cities located approximately 90 km downstream. To prevent the Spirit Lake blockage from ever being breached by overflow, the level of the lake surface was stabilized by the U.S. Army Corps of Engineers (USACE) at a safe level, first by pumping water over the potentially unstable natural dam in pipes using diesel pumps mounted on barges, and thereafter by draining lake water through a 3.3-m-diameter outlet tunnel that was bored 2.5 km through an adjacent bedrock ridge to form a permanent gravity drain that was completed in 1985 (Figure6). The USACE stabilized the outlets from two other debris-dammed lakes at Mount St. Helens (Coldwater and Castle Lakes) by constructing engineered outlet channels. The Spirit Lake drainage tunnel continues to function well, although periodic inspection and maintenance of the tunnel are necessary. None of the stabilized lakes at Mount St. Helens have had outbreaks (Sager and Budai[1989]; *Willingham*[2005]).

Lahar Diversion

Lahars can be prevented from spreading out and depositing in critical areas by keeping them channelized in modified natural channels or by engineering new channels. Such artificial channels (Figure 7a) must be sufficiently smooth, steep, and narrow (to maintain sufficient flow depth) in order to prevent in-channel deposition. The goal of such channelization is to keep lahars flowing so that they bypass critical areas. The effectiveness of this approach depends on lahar size and composition, channel dimensions, and construction techniques. Highly concentrated lahars (debris flows) can transport large boulders at high velocity and are extremely erosive, so channel bottoms and sides must be lined with concrete or stone masonry surfaces. Even so, hardened diversion channels may require frequent maintenance. Without hardening, lahars in diversion channels can easily erode channel boundaries and establish new flow paths. Channelization of lahar-prone streams draining volcanoes is relatively common in Japan and Indonesia (Smart [1981]; Japan Sabo Assoc. [1988]; Chanson [2004]).

Figure 7: Types of lahar diversion structures. (A) Engineered channel reach in small river draining Sakurajima volcano in southern Japan, where channel is revetted with reinforced concrete and engineered to be as steep, narrow, and smooth as possible, in order to divert lahars away from a developed area. (B) Training dike revetted with steel sheet piles on the lower flank of Usu vol-

cano, Japan and designed to deflect lahars away from buildings and other infrastructure. USGS photos by TCP.

Deflection and diversion structures also can be employed to reroute or redirect lahars away from critical infrastructure or communities. Structures include (a) tunnels or ramps to direct flows under or over roads, railroads, and pipelines; (b) training dikes (also termed levees or bunds) oriented sub-parallel to flow paths to guide lahars past critical areas; and (c) deflection berms oriented at sharper angles to flow paths to force a major course alteration in a lahar (Baldwin et al. [1987]; Hungr et al. [1987]; Huebl and Fiebiger [2005]; Willingham [2005]). However, lahar diversion may cause additional problems (and political resistance) if the diversion requires the sacrifice of only marginally less valuable land. Diversion ramps and tunnels are more practical for relatively small flows, whereas training dikes and deflection berms can be scaled to address a range of lahar magnitudes.

Dikes and berms are constructed typically of locally derived earthen material, but to be effective, these structures must be revetted (armored) on surfaces exposed to highly erosive lahars (Figure 7b). Revetment can be accomplished with thick layers of poured-in-place reinforced concrete, heavy concrete blocks or forms, heavy stone masonry faces or walls, stacked gabions, or steel sheet piles; layers of unreinforced concrete only centimeters thick cannot withstand erosion by large lahars (e.g., Paguican et al. [2009]). However, if a well-revetted dike is overtopped, rapid erosion of the unarmored back side of the dike can quickly cause dike failure and breaching nontheless (Paguican et al. [2009]) (Case study 3). In Japan, where probably more of these structures are constructed than anywhere else in the world, a major design criterion is that their orientation should ideally be less than 45° to the expected attack angle of a lahar to minimize overtopping and erosional damage (Ohsumi Works Office [1995]). Sometimes emergency levees are constructed without revetments, but this usually results in unsatisfactory performance, sometimes with disastrous results (Case study 1).

Case Study 3. Lahar and Sediment Containment and Exclusion Structures

In the months following the May 18, 1980 eruption of Mount St. Helens (Washington, USA), the U.S. Army Corps of Engineers (USACE) built a rock-cored earthen sediment-retention structure (N-1 sediment dam) as a short-term emergency measure to try to hold back lahars and some of the volcanic sediment expected to wash downstream (Willingham[2005]). The structure had two spillways made of rock-filled gabions covered with concrete mortar; it was 1,860 m long and 13 m high, and was located approximately 28 km downstream of the volcano. Neither the upstream nor downstream face of the dam was revetted. Within a month of completion, one of the spillways was damaged by high flow. That spillway was repaired and resurfaced with roller-compacted concrete. In slightly more than a year, the N-1 debris basin filled with about 17 million m3of sediment, and the bed of the river aggraded nearly 10 meters. During the summer of 1981, the USACE excavated 7.4 million m3from the debris basin, but the river replaced that amount and added more during the following winter. The dam was overtopped and breached in quick succession by two events in early 1982—a major winter flood in February and an eruption-triggered, 10-million-m3lahar in March. Overtopping caused deep erosion of the downstream face of the dam at several points, which led to breaching. Even the reinforced, roller-compacted concrete spillways were scoured tens of centimeters, exposing ends of steel reinforcing bars that were abraded to dagger-like sharpness. The extensive damage to the dam and the limited capacity of the catch basin resulted in abandonment of the project (Pierson and Scott[1985]; Willingham[2005]).

Several years later, the USACE started construction of another larger sediment-containment dam (the Sediment Retention Structure or SRS), which was completed in 1989 and further modified in 2012 (Figure8 a). It was built 9 km downstream of the original N-1 structure. In addition to trapping fluvial sediment, it was also designed to intercept and contain a possible future lahar (estimated peak discharge up to 6000 m3/s) from a potential breakout from Castle Lake. The SRS is a concrete-faced (upstream face), rock-cored, earthen dam about 550 m long, 56 m high, 21 m wide at the crest, and has a 122-m-wide armored spillway; its upstream catch basin is 13 km2in area and was designed to hold back about 200 million m3of sediment (USACE—Portland District, unpublished data). By 2005, infilled sediment reached the level of the spillway, and river bed-load sediment began to pass through the spillway, even though the catch basin was filled only to 40% of estimated capacity. After 2005, only a fraction of the river's sediment load was being intercepted, so raising of the spillway by an additional 2.1 m was completed in 2012 and experiments are continuing to induce greater sediment deposition in the upstream basin. The SRS has performed an important function in preventing large amounts of sediment from reaching and filling a reach of the Cowlitz River farther downstream and thus preventing serious seasonal flooding in communities along that river. No attempt has yet been made to excavate and remove sediment from behind the SRS.

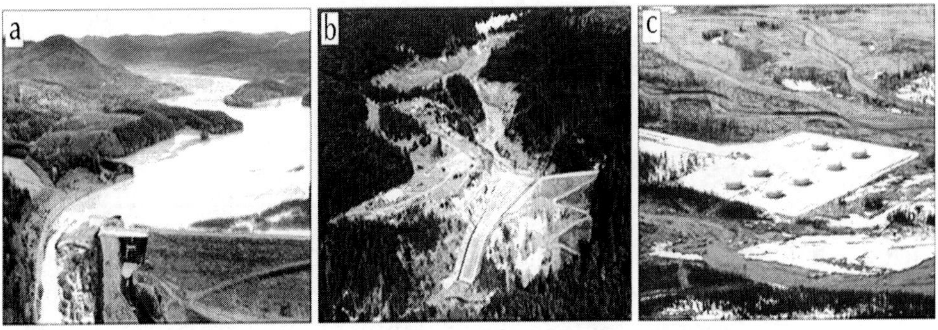

Figure 8: Examples of large-scale lahar containment and exclusion structures. (a) The Sediment Retention Structure (SRS) downstream of Mount St. Helens, USA, built specifically to contain potential lahars and eroded sediment (USGS photo by Adam Mosbrucker, 11 Nov 2012); the volcano is visible on the horizon on the left side of the image. (b) Mud Mountain Dam with a large concrete overflow spillway on the White River downstream of Mount Rainier (USA), (Stein [2001]). It was built as a flood-control structure but it also may function as a trap for at least part of future lahars because little water is normally impounded behind the dam (photo courtesy of U.S. Army Corps of Engineers). (c) Exclusion levees surrounding the Drift River oil terminal on an alluvial plain approximately 40 km downstream of Redoubt Volcano, Alaska (USGS photo by Chris Waythomas, 4 Apr 2009).

An example of a lahar exclusion structure is the levee system enclosing the Drift River Oil Terminal (DROT) in Alaska (USA), which is a cluster of seven oil storage tanks that receive crude oil from Cook Inlet oil wells via a pipeline, plus some buildings and an air strip (Dorava and Meyer[1994]; Waythomas et al.[2013]). The DROT is located on the broad, low-gradient flood plain at the mouth of the Drift River, about 40 km downstream of Redoubt Volcano (Figure8 c). Oil is pumped from these tanks to tankers anchored about 1.5 km offshore at a pumping-station platform. A U-shaped levee enclosure (built around the DROT but open at the downstream end) was raised to a height of 8 m following the 1989–1990 eruption, in order to increase protection of the facility from lahars and flooding. During both the 1989–1990 and 2009 eruptions of Redoubt, lahars were generated that flowed (at low velocity) up against the levees. Minor overtopping of the levees and backflow up from the open end caused some damage and periodic closure of the facility. The river bed aggraded to within 0.5 m of the levee crest in 2009, and the levees were thereafter reinforced and raised higher. The levee enclosure basically did its job, though it would have been more effective if the enclosure had been complete (on four sides).

Lahar Containment or Exclusion

Various structures can prevent lahars from reaching farther downstream, or seal off and protect critical areas while surrounding terrain is inundated. Sediment retention dams (Figure 8a) or containment dikes are used hold back as much sediment as possible but not necessarily water. To contain lahars, they must be constructed to withstand erosion and possible undercutting along their lateral margins and be tall enough to avoid overtopping. Under-design of these structures or inadequate removal of trapped sediment behind them can result in eventual overtopping and failure of the structure (e.g., Paguican et al. [2009]; Case study 3). The area upstream of a barrier where sediment is intended to accumulate is usually termed the catch basin or debris basin. Small excavated catch basins are also termed sand pockets. Such accumulation zones are typically designed to accommodate sediment from multiple flow events, and large tracts of land may be needed for this purpose. However, acquisition of land for this purpose can be problematic (Case study 1). If the design capacity is not large enough to accommodate all of the sediment expected to wash into a catch basin, provisions must be made to regularly excavate and remove accumulated sediment.

In addition to specially built lahar-related structures, pre-existing dams can sometimes be useful in containing all or most of the debris in a lahar (Figure 8b). Dams built for flood control or for impoundment of water for hydroelectric power generation or water supply can contain lahars and prevent them from reaching downstream areas, as long as (a) sufficient excess storage capacity exists behind the dam to accommodate the lahar volume, and (b) there is no danger of lahar-induced spillover at the dam in a way that could compromise dam integrity and lead to dam failure. Reservoir drawdown during volcanic activity might be necessary to ensure sufficient storage capacity to trap a lahar. This was done at Swift Reservoir on the south side of Mount St. Helens prior to the 1980 eruption, allowing it to successfully contain two lahars totaling about 14 million m^3 (Pierson [1985]).

Exclusion dikes can enclose and protect valuable infrastructure, as was done in 1989–1990 and 2009 to protect oil storage tanks at the mouth of the Drift River, Alaska, from lahars and volcanic floods originating from Redoubt Volcano (Dorava and Meyer [1994]; Waythomas et al. [2013]) (Case study 3; Figure 8c). Diked enclosures may be a more appropriate strategy than channelization, diversion, or deflection in areas with low relief where low channel gradients encourage lahar deposition and where areas to be protected are small relative to the amount of channelization or diking that otherwise would be required.

Check Dams to Control Lahar Discharge and Erosion

Some structures are built to slow down or weaken lahars as they flow down a channel. Check dams are low, ruggedly built dams that act as flow impediments in relatively steep stream channels (Figures 9 and 10). They have four functional roles: (a) to prevent or inhibit downcutting of the channel, which in turn inhibits erosion and entrainment of additional sediment; (b) to trap and retain some of a lahar's sediment, thereby decreasing its volume; (c) to add drop structures to the channel profile in order to dissipate energy and slow downstream progress of the lahar; and (d) to induce deposition in lower-gradient reaches between dams (Smart [1981]; Baldwin et al.[1987]; Hungr et al. [1987]; Johnson and McCuen [1989]; Armanini and Larcher [2001]; Chanson[2004]; Huebl and Fiebiger [2005]; deWolfe et al. [2008]).

Figure 9: Examples of permeable lahar flow-control structures. (a)Steel-pipe slit dam at Mount Unzen, Japan. (b) Drain-board screen at Mount Yakedake, Japan, after having stopped the bouldery head of a small debris-flow lahar. USGS photos by TCP.

Figure 10: Examples of impermeable lahar flow- and erosion-control structures. (a) Series of sheet-pile check dams with masonry aprons at Mount Usu, Japan. (b) Dam of rock-filled steel cribs at Mount Ontake, Japan. USGS photos by TCP.

Check dams are commonly built in arrays of tens to hundreds of closely spaced dams that give a channel a stair-step longitudinal profile.

Very low check dams are also called stepped weirs and are commonly constructed between larger check dams to act as hydraulic roughness elements for large flows (Chanson [2004]). A variety of styles and sizes of check dams have been developed, but fall into two basic categories: permeable or impermeable.

Permeable slit dams, debris racks, and open-grid dams (Figure 9a) are constructed of heavy tubular steel or structural steel beams, commonly with masonry bases and wing walls. Such structures are designed to act as coarse sieves, catching and retaining boulder-size sediment in a lahar but allowing finer material and water to pass through with depleted energy and mass. In addition to reducing the velocity of flows as they pass through, these dams also attenuate peak discharge. The effect is most pronounced on granular (clay-poor) debris-flow lahars that typically have steep, boulder-laden flow fronts. A variation on these vertically oriented structures is the drain-board screen (Azakami [1989]) (Figure 9b), which is a horizontally oriented steel grate or grill that performs the same sieving function for boulders as permeable dams when a lahar passes over the top of the grate, retaining coarse clasts while water and finer sediment drop down through the grate. Because of their orientation, these structures do not have to withstand the same high lateral forces as the upright permeable dams.

Impermeable check dams are composed of solid concrete, concrete with a packed earthen core, or steel cribs or gabion baskets filled with rocks and gravel (Figure 10). They may have small slits or pipes to allow exfiltration of water through the dam, in order to minimize impoundment of water. Gabions are used widely in the developing world because of their low construction costs—gravel fill often can be excavated locally from the channel bed, their permeability, and their flexibility, which can allow a dam to sag without complete failure if undermined by erosion. The crests of impermeable check dams commonly slope toward the center of the dam, where a notch or spillway is constructed, in order to direct streamflow or lahars over the dam onto a thick concrete apron extending downstream to protect the toe of the dam from erosion. Concrete sills or roughness elements commonly are placed at the downstream ends of aprons to further slow the flow that passes over the main dam. If upstream catch basins fill to capacity with sediment, check-dam functions are then limited to *a*, *c*, and *d* noted above, but full functionality can be restored if catch basins are regularly excavated.

Hazard Warning

Where communities already occupy lahar hazard zones or where transient populations move in and out, a lahar warning system can be an option that would allow an at-risk population to safely evacuate prior to lahar arrival, whether or not used in conjunction with engineered protection structures. Lahar warning systems can minimize fatalities, but they are not practical in every situation. In cases where populations are situated close to a lahar source area, there simply may be little or no time for a timely warning to be issued and for people to receive it in time to evacuate (Cardona [1997]; Pierson [1998]; Leonard et al. [2008]). Timing is even more challenging at volcanoes where lahars unrelated to ongoing or recent volcanic activity can occur—where volcanic edifices are weakened by hydrothermal alteration, for example, because lahar occurrence generally would not be anticipated. The decision of whether or not to install a warning system should also consider the long-term and ongoing needs for sustaining coordination and communication among the many organizations and individuals involved, regularly maintaining and testing the instrumentation, and keeping at-risk populations informed and prepared, especially where populations are transient.

Lahar warning systems have three basic components: (1) sensors or observers to detect an approaching lahar; (2) data acquisition, transmission, and evaluation systems to transfer and evaluate data to determine if there really is an approaching lahar; and (3) alert-notification systems to inform people that a lahar is coming. The spectrum of ways to accomplish these functions can range from simple 'low-tech' approaches largely involving human observers to more sophisticated 'high-tech' systems (Figure 11). In addition to these basic components that warn of an approaching lahar, integrated (often called "end-to-end") warning systems also include components that not only warn people but prepare them and lead them to respond proactively and to assume personal responsibility for evacuating. These additional components include pre-event planning and preparation; mechanisms to formulate and target appropriate warning messages; effective outreach to at-risk populations so that they understand what to do when a warning is received; establishment of evacuation routes and safe refuges that can be reached (generally on foot) before lahar arrival; and evacuation exercises with follow-up evaluation (Mileti and Sorenson[1990]; Basher [2006]; Leonard et al. [2008]).

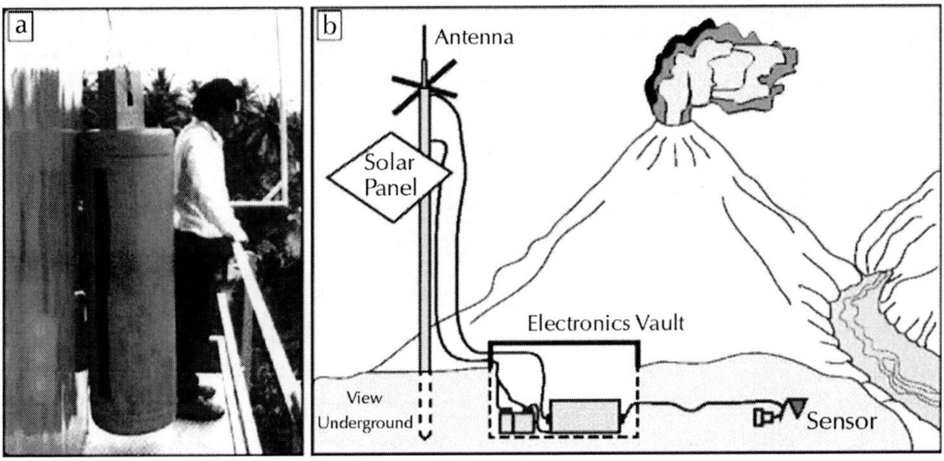

Figure 11: Examples of "low-tech" and "high-tech" lahar detection systems. (a) Human observer in lahar observation tower along a river that originates on Merapi volcano, Indonesia; observer strikes the large hanging steel drum ("tong-tong") with a steel bar after seeing or hearing an approaching lahar. USGS photo by TCP. (b) Schematic diagram of an acoustic flow monitor (AFM)—a sensor that detects ground vibrations generated by an approaching lahar, then telemeters that information in real time to a base station, where the signal is evaluated and a decision is made on whether or not to issue an alarm (see LaHusen [2005]).

Once a warning system becomes operational and depended upon, there must be sufficient ongoing funding and institutional commitment to continue operation indefinitely and to regularly educate and train the at-risk population. This is important because termination of a warning system while the hazard still exists may involve liability and ethical issues. Long-term operation costs include not only those for the normal maintenance of warning-system components, but also replacement costs if components are vandalized or stolen and, where necessary, costs for providing instrument-site security.

Volcano scientists play important roles, not only in developing or deploying warning system instrumentation, but also in training emergency managers to confidently interpret scientific and technical information from the monitoring systems. Scientists also can help to develop clear warning messages that are appropriate and understandable by affected populations (Mileti and Sorenson [1990]). Although lahar warning systems can issue false alarms, research shows that the "cry wolf" syndrome does not develop within affected populations as long as people understand the hazard and are later told about the possible reasons why a false warning was issued (Mileti and Sorenson [1990]; Haynes et al. [2008]).

'Low-Tech' Warning Systems

In some developing countries, effective low-tech warning systems employ human observers to alert threatened populations. Observers can be positioned at safe vantage points within view of lahar-prone river channels at times when flows have a high likelihood of occurring, such as during ongoing eruptions and during and following intense rainfall, particularly within the first few years after eruptions (de Bélizal et al. [2013]; Stone et al. [2014]). Observers stationed near lahar source areas are in a position to see or hear localized convection-cell rain storms that can trigger lahars, and human hearing can be very effective in detecting the approaching lahars themselves, often minutes before they come into view. The low-frequency rumbling sound caused by large boulders grinding against the river bed can carry hundreds or thousands of meters through the air and through the ground—a sound that is unmistakable to a trained observer. For example, a relatively small lahar occurring recently at Mount Shasta, California, sounded "like a freight train barreling down the canyon" and at times "like a thunder rumble" to a U.S. Forest Service climbing ranger (Barboza [2014]).

Once a lahar is detected, an observer can quickly issue an alert directly (by drum, siren, cellular phone, hand-held radio, etc.) to people living nearby (Figure 11a). This basic approach to lahar detection may be preferable where there is limited technical or financial capacity for maintaining sensors and other electronic equipment, where there are safe and accessible observation points, where there is high likelihood of expensive instruments being damaged or stolen without someone to guard them, where environmental conditions are challenging, or where electrical power and telecommunications are unreliable. Lahar detection by human observers is not immune to failure, however. Reliability is a function of the trustworthiness and alertness of the observers, their level of training, and the effectiveness of the alert notification method.

Automated Telemetered Warning Systems

Automated electronic warning systems can be used to detect approaching lahars and telemeter alerts in areas where electrical power, technical support capabilities, and funding are more assured. Systems also can be designed to detect anomalous rainfall or rapid snowmelt that could trigger lahars, sense incipient motion of an unstable rock mass or lake-impounding natural dam, or detect an eruption that could trigger a lahar (Marcial et al. [1996]; Sherburn and Bryan [1999]; LaHusen [2005]; Manville and Cronin [2007]; Leonard et al. [2008]; USGS [2013]) (Figure 11b). In order for data from any of these various sensors to be useful for alert notification, they must be transmitted from remote sites in real time to a receiving station. Transmission can be accomplished by either ground-based or satellite-based radio telemetry (LaHusen [2005]) or cellular phone (Liu and Chen [2003]). Alert notifications can occur either automatically when some threshold in the level of the detection signal is exceeded, or an intermediate step can involve emergency management personnel, who verify and validate the detection signal before an alert is issued. Coordination among multiple agencies is critical to the success of an automated system, because hardware and software development of the sensor and the data acquisition/transmission systems are typically handled by physical scientists and engineers, whereas the development, operation, and maintenance of warning systems are typically managed by emergency managers and law-enforcement personnel (Case study 4).

Case Study 4. The Mount Rainier Lahar Warning System

A significant volume of rock on the upper west flank of Mount Rainier (USA) has been extensively weakened (60–80% loss in unconfined strength) by hydrothermal alteration and is unstable (Watters et al.[2000]*; Finn et al.*[2001]*; John et al.*[2008]*). A lahar warning system was developed by the U.S. Geological Survey and Pierce County*

(Washington) to detect potential lahar initiation from this sector, and it was installed in 1995 by USGS and Pierce County personnel in the Carbon and Puyallup River valleys downstream of the weak and oversteepened rock mass (USGS[2013]*). The system is designed to warn tens of thousands of people who live in the downstream lahar hazard zone of an approaching lahar. Affected communities are situated from 40 to 80 km downstream of the volcano and could have from 12 minutes to 2 hours, depending on location, to evacuate after receiving a warning message. Since installation, the warning system has been maintained and operated by the Pierce County Department of Emergency Management, in collaboration with the Washington State Emergency Management Division.*

The system comprises specialized seismic sensors capable of detecting ground vibrations within a frequency range typical of lahars (30–80 Hz), a ground-based radio telemetry system for detection-signal transmission, and a combination of sirens, direct notification, and the Emergency Alert System (EAS) that utilizes NOAA weather radios for warning message dissemination (LaHusen[2005]*; USGS*[2013]*). County and state emergency-management agencies and city and county law-enforcement agencies collectively have responsibility for verifying and validating alerts from the sensors, activating warning sirens, and sending warning messages.*

Collaboration between all the agencies involved in lahar hazard warning and risk reduction at Mount Rainier is fostered by regular meetings of the "Mount Rainier Work Group". Such lahar warning systems require ongoing collaboration between scientists and emergency management officials, as well as regular maintenance and testing. Members of the at-risk population (including schools) have been assigned evacuation routes, have been informed about what to do when a warning message is received, and regularly participate in evacuation drills (Figure 3g).

Warning Message Development and Delivery

In the simplest warning systems, warning messages are delivered only as simple audible signals (drums, sirens, whistles, etc.), and the affected population must be informed beforehand about what the signals mean and what the appropriate response should be. In more sophisticated systems, incident-specific alert messages can be delivered to large populations simultaneously by cellular phone, the Internet, radio, or television. In these cases, the alert must convey a definitive and unambiguous message that effectively prompts individuals to take protective actions. Several factors influence the effectiveness of a warning message, including the content and style of the message, the type and number of dissemination channels, the number and pattern of warning statements, and the credibility of the warning source (Mileti and Sorenson [1990]).

Warning messages should be specific, consistent, certain, clear, and accurate (Mileti and Sorenson[1990]). To ensure credibility, message content should include a description of the hazard and how it poses a threat to people, guidance on what to do to maximize personal safety in the face of impending danger, location of the hazard, the amount of time people have to take action, and the source of the warning. The more specific a warning message is, the more likely the receiver is to accept the warning (Cola [1996]; Greene et al. [1981]). Emergency warnings without sufficient detail create information voids, and the affected population may then rely on ill-informed media commentators, friends, neighbors, or personal bias and perceptions to fill this void (Mileti and Sorenson [1990]). Input from volcano scientists is critical for some of this detail and specificity.

Both credibility and consistency of the warning message are important. At-risk populations commonly receive information from informal sources (for example, the media, friends, social media), sometimes more quickly than through various official channels during a crisis (Mileti[1999]; Leonard et al. [2008]; Dillman et al. [1982]; Mileti and Sorenson [1990]; Parker and Handmer [1998]; Mei et al. [2013]). For example, 40–60% of people in the vicinity of Mount St. Helens first received informal notification of the 1980 eruption (Perry and Greene [1983]; Perry[1985]). The proliferation of informal information channels today with the Internet and social media can benefit the warning dissemination process, because individuals are more likely to respond to a warning if it is confirmed by multiple sources (Cola [1996]; Mileti and Sorenson[1990]). But multiple sources become problematic if they advance conflicting information, causing individuals to become confused. Therefore, challenges for emergency managers and scientists are to keep reliable information flowing quickly and to maintain consistent messages, both during and after an emergency. Joint information centers can ensure that (a) there is consistency in official warning statements among multiple scientific and emergency-management agencies, (b) easy access is provided for the media to the official information and to experts who can explain it, and (c) the effectiveness of warning messages is monitored (Mileti and Sorenson [1990]; Driedger et al. [2008]).

Evacuation Training

Warnings are given so that people in a lahar flow path can move quickly out of harm's way. Sheltering in place is generally not a viable option. The lives of at-risk individuals may depend on understanding that they are living in, working in, driving through, or visiting a lahar hazard zone, as well as understanding what to do when they receive a warning (Mileti and Sorenson [1990]; Leonard, et al. [2008]). As the world witnessed in the 1985 Nevado del Ruiz disaster (Voight[1990]) (Case study 5), warnings that a lahar was bearing down on their town were not able to prevent catastrophic loss of life, because the warnings were issued without the population's understanding of the risk or how they should respond. To increase the likelihood of successful evacuations, scientists should encourage and help lead hazard-response exercises and evacuation drills, especially in areas with short time windows for evacuating hazard zones. These exercises and drills provide emergency managers the opportunity to identify weaknesses in the warning–evacuation process and to minimize potential delays that could result from confusion, insufficient information, or lack of understanding on what to do. They also provide scientists with a platform for discussing past catastrophes and the potential for future events. Holding an annual table-top exercise or community-wide evacuation drill on the anniversary of a past disaster can help to institutionalize and personalize the memory of past events, an important step if new community members are to take these threats seriously. A well-educated and trained community that possesses information about where they will get information and what emergency actions to take is less likely to be confused by warning messages, to resist evacuation orders, or to blame officials for ordering an evacuation when a catastrophic event fails to occur (e.g., Cardona [1997]). The goal for scientists and emergency managers is to create a "culture of safety" (cf., Wisner et al.[2004], p. 372) where at-risk individuals understand potential hazards, take personal responsibility for reducing their risks, understand how to respond to an event, and realize that lessening of risks requires actions from all levels of a community and government.

Case Study 5. The Nevado Del Ruiz Disaster

The 1985 Nevado del Ruiz lahar disaster, which cost approximately 21,000 lives in the town of Armero, Colombia (Figure1 a), is an excellent case study of the complexities that can lead to ineffective evacuation after warning messages are broadcast, poor emergency response, and a haphazard disaster recovery (Voight[1990]; *Hall*[1992]*). In post-event analyses, it was generally concluded that the Ruiz catastrophe was the result of cumulative human and bureaucratic errors, including lack of knowledge, misunderstanding and misjudgment of the hazard, indecision, and even political barriers to effective communication, rather than inadequate science or technical difficulties. Other factors contributing to the catastrophe included evacuation plans that had been prepared but not shared with the public, poorly equipped emergency management authorities, the absence of agreed-upon decision-making processes, and uncertainty about the pre-event hazard assessments that made public officials reluctant to issue an early evacuation order because of the potential economic and political costs. The hazard maps produced by scientists for Nevado del Ruiz prior to the eruption were highly accurate in their predictions of where lahars could go, but they were published only about a month before the disaster, giving little time for assimilation and responsive action by the emergency managers. Furthermore, production of the maps did not lead to effective risk communication, because the scientists who made the maps generally did not engage in conveying that risk information in understandable terms to officials and the public. Scientists may prepare excellent hazard assessments and maps, but unless they participate fully in conveying hazard information to officials and the public in ways that are understandable, disasters can still happen (Voight*[1990]; *Hall*[1992]*).*

Hazard Response and Recovery Planning

The first three risk-reduction strategies focus on minimizing losses through actions taken before a lahar occurs, but this fourth strategy determines the effectiveness of the immediate emergency response and the longer-term course of recovery after a lahar has occurred, which together define a community's resilience. Hazard response includes the rescue, emergency care, sheltering, and feeding of displaced persons, which is facilitated by a robust incident command system. Such a system could range from coordinated communication in a small village to a structured multi-agency protocol, such as NIMS (National Incident Management System) in the United States (FEMA[2014]). Recovery involves the reestablishment of permanent housing, infrastructure, essential services, and economic viability in the community.

Response to a lahar that has impacted a populated area can be difficult. Lahars present first responders, search-and-rescue teams, and disaster-management officials with challenges unlike some other disasters: (a) the area of impact can be extensive and locally covered by debris from crushed buildings and other structures; (b) the degree of impact is generally greatest toward the center of the impact zone and less along the edges; (c) lahars can transport victims and structures long distances from their initial locations; (d) survivors may be difficult to locate; (e) fresh lahar deposits commonly stay liquefied (like quicksand) for days to weeks, and upstream river flow may cut through a debris field, so that access to victims may be limited to hovering helicopters, small boats, or rescuers on the ground being confined to walking on logs or sheet of plywood (Figure 12); (f) once located, victims can be difficult to extract from the mud; and (g) critical facilities (hospitals, police and fire stations, etc.) may be inaccessible, damaged, or destroyed. These challenges can be critical, because the time window is small for getting injured victims to medical care, and uninjured victims trapped in liquefied mud can quickly become hypothermic. To minimize fatalities from a lahar, communities in hazard-prone areas should develop realistic rescue and response plans that are understood by all individuals and responsible agencies. In addition to developing search and rescue tactics, such plans should include identification of refuge zones, logistical resources, emergency social services, and security personnel that will be needed to establish emergency shelters and for survivors at those shelters, and for site access control and security (see UNDRO [1985], for an emergency plan example). Scientists can support emergency managers and public officials in the aftermath of a catastrophic event by assessing the likelihood of future lahars and floods, the suitability of areas for relief operations, and the evolving stability of lahar deposits.

Figure 12: Examples of challenges to rescue and recovery where thick liquefied mud and debris have flowed into a populated area—the Highway 530 (Oso, Washington) landslide disaster of 22 March 2014. Soft mud can preclude rescue of victims by responders on the ground, particularly in the first hours or days following a lahar. (a) Rescuer being lowered by helicopter to an area where ground is too soft to reach on foot (copyrighted AP photo by Dan Bates, used with permission). (b) Rescuer searching for victims using an inflatable boat, because flooding from backed-up river inundated part of the debris field (copyrighted AP photo by Elaine Thompson, used with permission).

Proper shelter planning is critical to minimize the potential for additional victims. Poor planning of emergency shelters and camps can create new disaster victims due to disease outbreaks and malnutrition if shelter is inadequate and timely supply of food, clean water, and medicine does not occur. Shelter planning should also take into account the quality of life and livelihoods for displaced populations. For example, 50 to 70% of people displaced by the 2010 eruption of Mt. Merapi (Indonesia) ignored evacuation orders and consistently returned (in some cases daily) to danger zones during the crisis because of the need to care for livestock and to check on possessions (Mei et al. [2013]). The lack of activities and work programs in the evacuation camps also can result in people leaving the shelters. In addition, if schools are used as shelters, then public education suffers because school buildings are occupied by evacuees. In countries with limited relief resources, people may be better served if extended families can temporarily house impacted relatives during emergencies. Community leaders, with assistance from scientists, can encourage residents to develop their own evacuation and relocation strategies.

Following an initial disaster response, recovery becomes the next goal. Restoring community functions is typically a top priority in the aftermath of an extreme event such as a lahar, but quick reconstruction may not be possible if key infrastructure, industrial parks, downtown cores of communities, and extensive areas of residential housing are buried or swept away (Tobin and Whiteford [2002]). Pre-event recovery planning, however, can allow resilient communities to recover more quickly by prioritizing the building of redundant and diversified back-up systems, services, and infrastructure into their communities beforehand. For transportation networks for example, this could mean having multiple routes to critical or essential facilities, predetermined appropriate sites for helipads or temporary airstrips, and storage sites for heavy equipment—all located outside of the hazard zone. Scientists can assist the development of recovery plans by providing advice on where future commercial, residential, and industrial districts could be located outside of hazard zones. A well thought-out recovery plan also provides an impacted community with opportunities for the established social fabric of a community to be maintained, for relocation to a safer site, and for comprehensive redevelopment that avoids haphazard or fragmented future growth.

Resettlement following a disaster is not simply a matter of rebuilding homes and infrastructure at a safer site. The quality of life, means of making a living, and social needs and networks of displaced populations must be recognized for resettlement to be successful, and residents must be part of the planning process. For example, Usamah and Haynes ([2012]) document low occupation rates of (and minimal owner investment in) government-provided housing at permanent relocation sites two years after the Mayon volcano (Philippines) eruption in 2006. They attribute this to the lack of community planning participation, lack of appreciation of original house design and function (for example, metal roofs on new houses make them hotter during the day than traditional houses with palm-thatch roofs), delays in utility infrastructure, no public facilities such as religious centers and schools, few livelihood options, and little long-term community development. Although authorities and donors (and residents) were satisfied that the new housing was safer, interviewees felt the long-term objective of facilitating sustainable lives was ignored. A similar reluctance to participate in a resettlement program was found at Colima volcano (Mexico) for many of the same reasons (Gavilanes-Ruiz et al. [2009]). Thus, community participation in long-term recovery planning is needed to ensure identification of the community's needs and the community's support.

Development of an effective recovery plan can ensure provision of a number of practical recovery needs. Those needs include: achievement of more appropriate land-use regulations, identification of funding sources for reconstruction, identification of resources and disposal sites for debris clearance, enlistment of economic support for recovering businesses, and adoption of new construction standards. Recovery plans help ensure that reconstruction after the event does not reoccupy a hazard zone or happen in an ad hoc fashion. Scientists can contribute to this planning process by (a) helping public officials visualize the probable physiographic, geologic, and hydrologic realities of a post-event landscape; and (b) identifying what post-event hazards would be relevant for the community.

SCIENTIST ROLES IN LAHAR RISK REDUCTION

All four of the basic strategies for lahar-hazard risk reduction—hazard avoidance, modification, warning, and response/recovery—require the input and judgment of volcano scientists, even though emergency managers and public officials have the responsibility for their planning and implementation. In addition, scientists play a critical role in educating emergency managers, public officials, and at-risk populations about lahar hazards. Specific ways that scientists can participate are discussed in the sections above.

Some scientists are uncomfortable participating in processes that are influenced (if not dominated) by social, economic, and political factors. However, risk managers cannot successfully manage natural threats to communities without involvement by scientists (Peterson [1988], [1996]; Hall[1992]; Haynes et al. [2008]). Peterson ([1988]) goes as far to say that scientists have an ethical obligation to effectively share their knowledge to benefit society by making their knowledge understandable to non-scientists. Scientists can communicate hazard information to the public through formal and informal face-to-face meetings, through public presentations, and through the media. Qualities exhibited by scientists that enhance their trustworthiness in the eyes of the public are reliability (consistency and dependability in what they say), competence (having the skills and ability to do the job), openness (having a relaxed, straightforward attitude and being able to mix well and become 'part of the community'), and integrity (having an impartial and independent stance) (Pielke [2007]; Haynes et al. [2008]). Yet there is always a potential for friction and other distractions during the stressful time of a volcano crisis, and scientists should recognize and try to avoid the various problems related to personal and institutional interactions that have plagued the credibility of scientists during past volcanic crisis responses, such as communications breakdowns and disputes among scientists (with different messages coming from different scientists), scientists advocating for particular mitigation strategies, scientists avoiding or "talking down" to the public, poor scientific leadership, failure to recognize cultural differences between themselves and affected populations, and failure to share information and scarce resources (Newhall et al. [1999]).

Effective lahar-hazard risk reduction cannot occur unless the hazard and its attendant risks are recognized by authorities and the public, and this recognition is affected by the willingness and ability of scientists to communicate hazards information (Peterson [1988]). The contributions of scientists will be effective if they are willing to embrace their educational, interpretive, and advisory roles, to work in partnership with officials and the public, and to be sensitive to the cultural norms of the society in which they are working. Scientists must be willing and able to participate in community events, hone skills related to public speaking, work with the media, and work one-on-one with community leaders. As Newhall et al. ([1999]) state, the guiding principle for scientists during volcanic crises should be to promote public safety and welfare. This principle extends to non-crisis situations, as well, and scientists can and should work with officials and the public frequently to lessen the risk from future lahars. In short, lahar-hazard risk reduction cannot be effectively accomplished without the active, impartial involvement of qualified scientists.

AUTHORS' CONTRIBUTIONS

TCP developed the risk-reduction strategy categorization, evaluated the effectiveness of many of the strategies, and wrote approximately 60% of the manuscript. NJW summarized the ways that community vulnerability and risk reduction can be applied to lahar hazards, and he wrote approximately 40% of the manuscript. CLD edited and revised an early draft of the section on hazard and risk education. This section is in large part a distillation of CLD's findings on how hazards information can be communicated effectively and understandably. All authors read and approved the final manuscript.

ACKNOWLEDGMENTS

This review of strategies for lahar risk reduction is based not only on the literature cited, but also on observations made by the authors of the practical application of these techniques in many parts of the world, combined with their own direct experience and research. Photographs with credits in the form of initials were taken by the authors. Work by the authors on this topic has been supported over the years by the USGS Volcano Hazards Program, the USGS/USAID–OFDA Volcano Disaster Assistance Program, and the USGS Land Change Science Program. We thank Kelvin Rodolfo, Franck Lavigne, and one anonymous reviewer for their insightful reviews of an earlier version of the article. Any use of trade, product, or firm names is for descriptive purposes only and does not imply endorsement by the U.S. Government.

REFERENCES

1. Armanini A, Larcher M (2001) Rational criterion for designing opening of slit check dam. J Hydraul Eng ASCE 127:94-104
2. Aucker M, Sparks R, Siebert L, Crosweller H, Ewert J (2013) A statistical analysis of the global historical volcano fatalities record. J Appl Volcanol 2(2):24
3. Azakami S (1989) Recent trends in sabo control measures in Japan. In: Proceedings of the International Symposium on Erosion

and Volcanic Debris Flow Technology. Ministry of Public Works, Yogyakarta, Indonesia. pp KO2-1-KO2-9

4. Baldwin JE, Donley HF, Howard TR (1987) On debris flow/ avalanche mitigation and control, San Francisco Bay area, California. In: Costa JE, Wieczorek GF (eds) Debris Flows/ Avalanches: Process, Recognition, and Mitigation, Geol Soc Am Reviews in Engineering Geology, pp. 223-236
5. Barboza T (2014) Forest Service Thinks California's Drought Caused a Massive Mudslide. Los Angeles Times (Science section).
6. Basher R (2006) Global early warning systems for natural hazards: systematic and people-centred. Phil Trans Royal Soc A 364:2167-2182
7. Beyers JL (2004) Postfire seeding for erosion control: effectiveness and impacts on native plant communities. Conserv Biol 18:947-956
8. Bird D, Gisladottir G, Dominey-Howes D (2010) Volcanic risk and tourism in southern Iceland: Implications for hazard, risk and emergency response education and training. J Volcanol Geotherm Res 18:33-48
9. Cardona O (1997) Management of the volcanic crises of Galeras volcano: social, economic and institutional aspects. J Volcanol Geotherm Res 77:313-324
10. Chanson H (2004) Sabo check dams—mountain protection systems in Japan. Internat J River Basin Mgmt 2:301-307
11. Chenet M, Grancher D, Redon M (2014) Main issues of an evacuation in case of volcanic crisis: social stakes in Guadeloupe (Lesser Antilles Arc). Nat Hazards 73:2127-2147
12. Chester D (2005) Theology and disaster studies: the need for dialogue. J Volcanol Geotherm Res 146:319-328
13. Cola R (1996) Responses of Pampanga households to lahar warnings--lessons from two villages in the Pasig-Potrero River watershed. In: Newhall CG, Punongbayan RS (eds) Fire and Mud: Eruptions and Lahars of Mount Pinatubo, Philippines, University of Washington Press, Seattle, WA. pp 141-149
14. Collins BD, Dunne T (1986) Erosion of tephra from the 1980 eruption of Mount St. Helens. Geol Soc Am Bull 97:896-905

15. Crittenden K (2001) Can this town survive? Case study of a buried Philippine town. Nat Hazards Rev 2:72-79
16. Crittenden K, Rodolfo K (2002) Bacolor Town and Pinatubo Volcano, Philippines: coping with recurrent lahar disaster. In: Torrence R, Grattan J (eds) The Archaeology of Natural Disasters, One World Archaeology Series, Routledge, London. pp 43-65
17. Cronin S, Petterson M, Taylor P, Biliki R (2004) Maximising multi-stakeholder participation in government and community volcanic hazard management programs: a case study from Savo, Solomon Islands. Nat Hazards 33:105-136
18. Cronin SJ, Gaylord DR, Charley D, Alloway BV, Wallez S, Esau JW (2004) Participatory methods of incorporating scientific with traditional knowledge for volcanic hazard management on Ambae Island, Vanuatu. Bull Volcanol 66:652-668
19. Cummans J (1981) Chronology of mudflows in the South Fork and North Fork Toutle River following the May 18 eruption. In: Lipman PW, Mullineaux DR (eds) The 1980 eruptions of Mount St, US Geol Surv Professional Paper 1250, Helens, Washington. pp 479-486
20. de Bélizal E, Lavigne F, Robin AK, Sri Hadmoko D, Cholik N, Thouret JC, Sawudi DS, Muzani M, Sartohadi J, Vidal C (2013) Rain-triggered lahars following the 2010 eruption of Merapi volcano, Indonesia: A major risk. J Volcanol Geotherm Res 261:330-347
21. deWolfe VG, Santi PM, Ey J, Gartner JE (2008) Effective mitigation of debris flows at Lemon Dam, La Plata County, Colorado. Geomorphol 96:366-377
22. Dillman DA, Schwalbe ML, Short JF (1982) Communication behavior and social impacts following the May 18, 1980 eruption of Mount St. Helens. In: Keller SAC (ed) , Three Years Later, Pullman, WA, Washington State University, Mt. St. Helens. pp 173-179
23. Dominey-Howes D, Minos-Minopoulos D (2004) Perceptions of hazard and risk on Santorini. J Volcanol Geotherm Res 137:285-310
24. Dorava JM, Meyer DF (1994) Hydrologic hazards in the lower Drift River basin associated with the 1989–1990 eruption of

Redoubt Volcano, Alaska. In: Miller TP, Chouet BA (eds) The 1989–1990 Eruptions of Redoubt Volcano, Alaska, pp. 387-407

25. Driedger CL, Scott WE (2008) Mount Rainier—Living Safely With a Volcano in your Backyard.
26. Driedger CL, Scott WE (2010) Volcano hazards. In: Schelling J, Nelson D (eds) Media guidebook for natural hazards in Washington—addressing the threats of tsunamis and volcanoes, Washington Military Department Emergency Management Division,
27. Driedger CL, Wolfe EW, Scott KM (1998) Living With a Volcano in your Back Yard: Mount Rainier Volcanic Hazards—A Prepared Presentation for use by Public Officials and Educators.
28. Driedger C, Stout T, Hawk J (2002) The Mountain is a Volcano!—Addressing geohazards at Mount Rainier. J Assoc Nat Park Rangers 18:14-15
29. Driedger C, Doherty A, Dixon C, Faust L (2014) Living with a volcano in your backyard—An educator's guide with emphasis on Mount Rainier (ver. 2.0, December 2014).
30. Driedger CL, Neal CA, Knappenberger TH, Needham DH, Harper RB, Steele WP (2008) Hazard information management during the autumn 2004 reawakening of Mount St. Helens volcano, Washington. In: Sherrod DR, Scott WE, Stauffer PH (eds) A Volcano Rekindled: The Renewed Eruption of Mount St. Helens, 2004–2006. US Geol Surv Professional Paper 1750, pp. 505-519
31. Driedger CL, Westby L, Faust L, Frenzen P, Bennett J, Clynne M (2010) 30 cool facts about Mount St. Helens.
32. Dzurisin D, Driedger CL, Faust L (2013) Mount St. Helens, 1980 to Now—What's Going On? US Geol Surv Fact Sheet 2013–3014.
33. (2014) National Incident Management System. U.S. Federal Emergency Management Agency.
34. Finn CA, Sisson TW, Deszcz-Pan M (2001) Aerogeophysical measurement of collapse-prone hydrothermally altered zones at Mount Rainier volcano. Nature 409:600-603
35. Frenzen PM, Matarrese MT (2008) Managing public and media response to a reawakening volcano: lessons from the 2004 eruptive activity of Mount St. Helens. In: Sherrod DR, Scott WE,

Stauffer PH (eds) A Volcano Rekindled: The Renewed Eruption of Mount St. Helens, 2004–2006. US Geol Surv Professional Paper 1750, pp. 493-503

36. Gaillard J, Maceda A (2009) Participatory three-dimensional mapping for disaster risk reduction. Participatory Learn Action 60:109-118
37. Gardner CA, Guffanti MC (2006) US Geological Survey's Alert Notification System for Volcanic Activity.
38. Gardner CA, Scott WE, Major JJ, Pierson TC (2000) Mount Hood—History and Hazards of Oregon's Most Recently Active Volcano.
39. Gavilanes-Ruiz JC, Cuevas-Muñiz A, Varley N, Gwynne G, Stevenson J, Saucedo-Girón R, Pérez-Pérez A, Aboukhalil M, Cortés-Cortés A (2009) Exploring the factors that influence the perception of risk: The case of Volcán de Colima, Mexico. J Volcanol Geotherm Res 186:238-252
40. Gray DH, Sotir RB (1996) Biotechnical and Soil Bioengineering Slope Stabilization: a Practical Guide for Erosion Control. Wiley-Interscience, New York.
41. Greene M, Perry R, Lindell M (1981) The March 1980 eruptions of Mt. St. Helens: citizens' perceptions of volcano threat. Disasters 5:49-66
42. Gregg C, Houghton B, Johnston D, Paton D, Swanson D (2004) The perception of volcanic risk in Kona communities from Mauna Loa and Hualalai volcanoes, Hawai'i. J Volcanol Geotherm Res 130:179-196
43. Guadagno FM, Revellino P (2005) Debris avalanches and debris flows of the Campania Region, southern Italy (Chapter 19). In: Jakob M, Hungr O (eds) Debris-flow Hazards and Related Phenomena, Praxis/Springer, Berlin. pp 489-518
44. Guffanti M, Brantley SR, Cervelli PF, Nye CJ, Serafino GN, Siebert L, Venezky DY, Wald L (2007) Technical-information products for a national volcano early warning system.
45. Hall M (1992) The 1985 Nevado del Ruiz eruption—scientific, social, and governmental response and interaction before the event, Chapter 6. In: McCall G, Laming D, Scott S (eds) Geohazards—Natural and Man-Made, Chapman and Hall, London. pp 43-52

46. Haynes K, Barclay J, Pidgeon N (2007) Volcanic hazard communication using maps: an evaluation of their effectiveness. Bull Volcanol 70:123-138
47. Haynes K, Barclay J, Pidgeon N (2008) The issue of trust and its influence on risk communication during a volcanic crisis. Bull Volcanol 70:605-621
48. Hicks A, Simmons P, Loughin S (2014) An interdisciplinary approach to volcanic risk reduction under conditions of uncertainty. Nat Hazards Earth Syst Sci 14:1871-1887
49. Hoblitt RP, Walder JS, Driedger CL, Scott KM, Pringle PT, Vallance JW (1998) Volcano hazards from Mount Rainier, Washington, Revised 1998.
50. Holtz RD, Schuster RL (1996) Stabilization of slopes. In: Turner AK, Schuster RL (eds) Landslides: Investigation and Mitigation, pp. 439-473
51. Huebl J, Fiebiger G (2005) Debris-flow mitigation measures (Chapter 18). In: Jakob M, Hungr O (eds) Debris-flow Hazards and Related Phenomena, Praxis/Springer, Berlin. pp 445-487
52. Hungr O, Morgan GC, VanDine DF, Lister DR (1987) Debris flow defenses in British Columbia. In: Costa JE, Wieczorek GF (eds) Debris Flows/Avalanches: Process, Recognition, and Mitigation, pp. 201-222
53. (1995) Understanding Volcano Hazards (video).
54. (1996) Reducing Volcanic Risk (video).
55. Janda RJ, Daag AS, Delos Reyes PJ, Newhall CG, Pierson TC, Punongbayan RS, Rodolfo KS, Solidum RU, Umbal JV (1996) Assessment and response to lahar hazard around Mount Pinatubo, 1991 to 1993. In: Newhall CG, Punongbayan RS (eds) Fire and Mud: Eruptions and Lahars of Mount Pinatubo, Philippines, Philippine Institute of Volcanology and Seismology, Quezon City, and University of Washington Press, Seattle. pp 107-139
56. (1988) Sabo works on active volcanoes in Japan. Japan Sabo Assoc, Tokyo.
57. John DA, Sisson TW, Breit GN, Rye RO, Vallance JW (2008) Characteristics, extent and origin of hydrothermal alteration at Mount Rainier Volcano, Cascades Arc, USA: Implications for

debris-flow hazards and mineral deposits. J Volcanol Geotherm Res 175:289-314
58. Johnson PA, McCuen RH (1989) Slit dam design for debris flow mitigation. J Hydraul Eng ASCE 115:1293-1296
59. Lagmay AMF, Rodolfo KS, Siringan FP, Uy H, Remotigue C, Zamora P, Lapus M, Rodolfo R, Ong J (2007) Geology and hazard implications of the Maraunot notch in the Pinatubo Caldera, Philippines. Bull Volcanol 69:707-809
60. LaHusen R (2005) Debris-flow instrumentation (Chapter 12). In: Jakob M, Hungr O (eds) Debris-flow hazards and related phenomena, Praxis-Springer, Berlin. pp 291-304
61. Lane LR, Tobin GA, Whiteford LM (2003) Volcanic hazard or economic destitution: hard choices in Baños, Ecuador. Environ Haz 5(1–2):23-34
62. Lavigne F, De Coster B, Juvin N, Flohic F, Gaillard J-C, Texier P, Morin J, Sartohadi J (2008) People's behaviour in the face of volcanic hazards: Perspectives from Javanese communities, Indonesia. J Volcanol Geotherm Res 172(3–4):273-287
63. Leonard GS, Johnston DM, Paton D, Christianson A, Becker J, Keys H (2008) Developing effective warning systems: Ongoing research at Ruapehu volcano, New Zealand. J Volcanol Geotherm Res 172:199-215
64. Liu KF, Chen SC (2003) Integrated debris-flow monitoring system and virtual center. In: Rickenmann F, Chen CL (eds) Debris-Flow Hazards Mitigation: Mechanics, Prediction, and Assessment (Proceedings of the Third International Conference in Davos, Switzerland), Mill press, Rotterdam. pp 767-774
65. Loughlin SC, Baxter PJ, Aspinall WP, Darroux B, Harford CL, Miller AD (2002) Eyewitness accounts of the 25 June 1997 pyroclastic flows and surges at Soufrière Hills Volcano, Montserrat, and implications for disaster mitigation. Geol Soc London, Memoir 21:211-230
66. Major JJ, Janda RJ, Daag AS (1996) Watershed disturbance and lahars on the east side of Mount Pinatubo during the mid-June 1991 eruptions. In: Newhall CG, Punongbayan RS (eds) Fire and Mud: Eruptions and Lahars of Mount Pinatubo, Philippines, Philippine Institute of Volcanology and Seismology, Quezon City, and University of Washington Press, Seattle. pp 895-919

67. Manville VR (2004) Paleohydraulic analysis of the 1953 Tangiwai lahar: New Zealand's worst disaster. Acta Vulcanol 16:137-152

68. Manville VR, Cronin SJ (2007) Breakout lahar from New Zealand's crater lake. EOS Trans Am Geophys Union 88(43):441-442

69. Manville VR, White JDL, Houghton BF, Wilson CJN (1999) Paleohydrology and sedimentology of a post-1.8 ka breakout flood from intracaldera Lake Taupo, North Island, New Zealand. Geol Soc Am Bull 111:1435-1447

70. Marcial S, Melosantos AA, Hadley KC, LaHusen RG, Marso JN (1996) Instrumental lahar monitoring at Mount Pinatubo. In: Newhall CG, Punongbayan RS (eds) Fire and Mud: Eruptions and Lahars of Mount Pinatubo, Philippines, University of Washington Press, Seattle, WA. pp 1015-1022

71. McGuire WJ, Solana MC, Kilburn CRJ, Sanderson D (2009) Improving communication during volcanic crises on small, vulnerable islands. J Volcanol Geotherm Res 183:63-75

72. Mei ETW, Lavigne F, Picquout A, de Bélizal E, Brunstein D, Grancher D, Sartohadi J, Cholik N, Vidal C (2013) Lessons learned from the 2010 evacuations at Merapi volcano. J Volcanol Geotherm Res 261:348-365

73. Michel-Kerjan EO (2010) Catastrophe economics: the National Flood Insurance Program. J Econ Perspect 24:165-186

74. Mileti DS (1999) Disasters by Design: A Reassessment of Natural Hazards in the United States. Joseph Henry Press, Washington, DC.

75. Mileti DS, Sorenson JH (1990) Communication of Emergency Public Warnings: A Social Science Perspective and State-of-the-Art Assessment. Report ORNL-6609. Oak Ridge National Laboratory, Oak Ridge, TN.

76. (1995) Slope Stabilization and Erosion Control—A Bioengineering Approach. Chapman and Hall, London.

77. Mothes PA, Hall ML, Janda RJ (1998) The enormous Chillos Valley lahar: an ash-flow-generated debris flow from Cotopaxi Volcano, Ecuador. Bull Volcanol 59:233-244

78. Myers B, Driedger CL (2008) Eruptions in the Cascade Range during the past 4,000 years: US Geol Surv General Information Product 63 (poster).

79. Myers B, Driedger CL (2008) Geologic Hazards at Volcanoes.
80. Némath K, Cronin SJ (2009) Volcanic structures and oral traditions of volcanism of Western Samoa (SW Pacific) and their implications for hazard education. J Volcanol Geotherm Res 186:223-237
81. Neumann van Padang M (1960) Measures taken by the authorities of the Vulcanological Survey to safeguard the population from the consequences of volcanic outbursts. Bull Volcanol 23(2):181-193
82. Newhall CG, Punongbayan RS (1996) The narrow margin of successful volcanic-risk mitigation. In: Scarpa R, Tilling RI (eds) Monitoring and Mitigation of Volcano Hazards, Springer, Berlin. pp 807-838
83. Newhall C, Aramaki S, Barberi F, Blong R, Calvache M, Cheminee J, Punongbayan R, Siebe C, Simkin T, Sparks S, Tjetjep W (1999) Professional conduct of scientists during volcanic crises. Bull Volcanol 60:323-334
84. (1995) Debris flow at Sakurajima, 2. Kyushu Regional Construction Bureau, Ministry of Construction, Tokyo.
85. O'Shea BE (1954) Ruapehu and the Tangiwai disaster. NZ J Sci Tech B36:174-189
86. Paguican EMR, Lagmay AMF, Rodolfo KS, Rodolfo RS, Tengonciang AMP, Lapus MR, Balisatan EG, Obille EC Jr (2009) Extreme rainfall-induced lahars and dike breaching, 30 November 2006, Mayon Volcano, Philippines. Bull Volcanol 71:845-857
87. Parker DJ, Handmer JW (1998) The role of unofficial flood warning systems. J Contingencies Crisis Manag 6:45-60
88. Paton D, Millar M, Johnston D (2001) Community resilience to volcanic hazard consequences. Nat Hazards 24:157-169
89. Paton D, Smith L, Daly M, Johnston D (2008) Risk perception and volcanic hazard -mitigation: Individual and social perspectives. J Volcanol Geotherm Res 172:179-188
90. Perry RW (1985) Comprehensive Emergency Management: Evacuating Threatened Populations. JAI Press, Greenwich, CT.
91. Perry RW, Greene MR (1983) Citizen Response to Volcanic Eruptions: The Case of Mt. St. Helens. Irvington Publishers, New York, NY.

92. Peterson DW (1988) Volcanic hazards and public response. J Geophys Res 93(B5):4161-4170
93. Peterson DW (1996) Mitigation measures and preparedness plans for volcanic emergencies. In: Scarpa R, Tilling RI (eds) Monitoring and Mitigation of Volcano Hazards, Springer, Berlin. pp 701-718
94. Pielke RA Jr (1999) Nine fallacies of floods. Clim Chang 42:413-438
95. Pielke RA Jr (2007) The Honest Broker—Making Sense of Science in Policy and Politics. Cambridge University Press, Cambridge.
96. (2008) Mount Rainier Volcanic Hazards Plan DEM. Pierce County (Washington) Dept. of Emergency Management (working draft).
97. (2014) Volcanic hazard areas, Chapter 18E.60, Title 18E, Pierce County (Washington) Code, adopted 2004—Ordinance No. 2004–57s, 2014 edition.
98. Pierson TC (1985) Initiation and flow behavior of the 1980 Pine Creek and Muddy River lahars, Mount St. Helens, Washington. Geol Soc Am Bull 96:1056-1069
99. Pierson TC (1989) Hazardous hydrologic consequences of volcanic eruptions and goals for mitigative action -- an overview. In: Starosolszky O, Melder OM (eds) Hydrology of Disasters. Proc. World Meteorological Organization Technical Conference, Geneva, November, 1988, James and James, London. pp 220-236
100. Pierson TC (1998) An empirical method for estimating travel times for wet volcanic mass flows. Bull Volcanol 60:98-109
101. Pierson TC, Scott KM (1985) Downstream dilution of a lahar: Transition from debris flow to hyperconcentrated streamflow. Water Resour Res 21:1511-1524
102. Pierson TC, Janda RJ, Thouret JC, Borrero CA (1990) Perturbation and melting of snow and ice by the 13 November 1985 eruption of Nevado del Ruiz, Colombia, and consequent mobilization, flow, and deposition of lahars. J Volcanol Geotherm Res 41:17-66
103. Pierson TC, Major JJ, Amigo A, Moreno H (2013) Acute sedimentation response to rainfall following the explosive phase of the 2008–2009 eruption of Chaitén volcano, Chile. Bull Volcanol 75:723 doi:10.1007/s00445-013-0723-4

104. Prater CS, Lindell MK (2000) Politics of hazard mitigation. Nat Hazards Rev 1:73-82
105. Rodolfo K (1995) Pinatubo and the Politics of Lahar—Eruption and Aftermath, 1991. University of the Philippines Press, Manila.
106. Ronan K, Johnston D (2005) Promoting Community Resilience in Disasters: The Role for Schools, Youth, and Families. Springer, New York.
107. Ronan K, Paton D, Johnston D, Houghton B (2000) Managing societal uncertainty in volcanic hazards--a multidisciplinary approach. Disast Prev Mgmt 9:339-348
108. Sager JW, Budai CM (1989) Geology and construction of the Spirit Lake outlet tunnel, Mount St. Helens, Washington. In: Galster RW (ed) Engineering Geology in Washington, Washington State Dept of Natural Resources Bull 78, Olympia, Washington. pp 1229-1234
109. Sager JW, Chambers DR (1986) Design and construction of the Spirit Lake outlet tunnel, Mount St. Helens, Washington. In: Schuster RL (ed) Landslide Dams—Processes, Risk, and Mitigation, pp. 42-58
110. Schelling J, Prado L, Driedger C, Faust L, Lovellford P, Norman D, Schroedel R, Walsh T, Westby L (2014) Mount Rainier is an active volcano--are you ready for an eruption?.
111. Schiechtl HM, Stern R (1996) Ground Bioengineering Techniques for Slope Protection and Erosion Control. Blackwell Scientific, Oxford.
112. Scott KM (1988) Origin, behavior, and sedimentology of prehistoric catastrophic lahars at Mount St. Helens, Washington. In: Clifton HE (ed) Sedimentologic Consequences of Convulsive Geologic Events, pp. 23-36
113. Scott W, Pierson T, Schilling S, Costa J, Gardner C, Vallance J, Major J (1997) Volcano hazards in the Mount Hood region, Oregon.
114. Scott KM, Macías JL, Naranjo JA, Rodríguez S, McGeehin JP (2001) Catastrophic debris flows transformed from landslides in volcanic terrains: Mobility, hazard assessment, and mitigation strategies.

115. Sherburn S, Bryan CJ (1999) The eruption detection system: Mt. Ruapehu, New Zealand. Seismol Res Lett 70:505-511

116. Smart GM (1981) Volcanic debris control, Gunung Kelud, East Java. In: Davies TRH, Pearce AJ (eds) Proceedings of Erosion and Sediment Transport in Pacific Rim Steeplands Symposium, Internat Assoc of Hydrol Sci Pub 132, Christchurch, N.Z. pp 604-623

117. Stein AJ (2001) Mud Mountain Dam. The Online Encyclopedia of Washington State History (HistoryLink.org Essay 3584). [http://www.historylink.org/essays/output.cfm?file_id=3584]*webcite* http://www.historylink.org/essays/output.cfm?file_id=3584 . Accessed 18 Feb 2014

118. Stone J, Barclay J, Simmons P, Cole PD, Loughlin SC, Ramón P, Mothes P (2014) Risk reduction through community-based monitoring: the *vigías* of Tungurahua, Ecuador. J Appl Volcanol 3:11 (online publication preview)

119. Suryo I, Clarke MCG (1985) The occurrence and mitigation of volcanic hazards in Indonesia and exemplified at the Mount Merapi, Mount Kelut and Mount Galunggung volcanoes. Quart J Eng Geol 18:79-98

120. Tayag J, Punongbayan R (1994) Volcanic disaster mitigation in the Philippines—experience from Mt. Pinatubo Disast 18:1-15

121. Theissen MS (1992) The role of geosynthetics in erosion and sediment control. Geotext Geomembr 11:535-550

122. Tobin GA (1995) The levee love affair—a stormy relationship? Water Resour Bull (JAWRA, J Am Water Resour Assoc) 31(3):359-367

123. Tobin G, Whiteford L (2002) Community resilience and volcano hazard: The eruption of Tungurahua and evacuation of the Faldas in Ecuador. Disasters 26:28-48

124. Umbal JV, Rodolfo KS (1996) The 1991 lahars of southwestern Mount Pinatubo and evolution of the lahar-dammed Mapanuepe Lake. In: Newhall CG, Punongbayan RS (eds) Fire and Mud: Eruptions and Lahars of Mount Pinatubo, Seattle, WA, University of Washington Press, Philippines. pp 951-970

125. (1985) Volcanic emergency management. United Nations, Office of the Disaster Relief Coordinator, New York.

126. Usamah M, Haynes K (2012) An examination of the resettlement program at Mayon Volcano: what can we learn from sustainable volcanic risk reduction? Bull Volcanol 74:839-859
127. (1996) Perilous Beauty—The Hidden Dangers of Mount Rainier (video).
128. (1998) At Risk: Volcano Hazards from Mount Hood, Oregon (video).
129. (2010) Mount St. Helens: A Catalyst for Change. Video program, 6 min 46 sec.
130. (2013) Monitoring Lahars at Mount Rainier.
131. Valentin C, Poesen J, Li Y (2005) Gully erosion: Impacts, factors, and control. Catena 63:132-153
132. Vallance JW (2000) Lahars. In: Sigurdsson H, Houghton BF, McNutt SR, Rymer H, Stix J (eds) Encyclopedia of Volcanoes, Academic Press, San Diego. pp 601-616
133. Vallance JW, Scott KM (1997) The Osceola Mudflow from Mount Rainier: Sedimentology and hazard implications of a huge clay-rich debris flow. Geol Soc Am Bull 109:143-163
134. Voight B (1990) The 1985 Nevado del Ruiz volcano catastrophe: Anatomy and retrospection. J Volcanol Geotherm Res 44:349-386
135. Voight B (1996) The management of volcano emergencies: Nevado de Ruiz. In: Scarpa R, Tilling RI (eds) Monitoring and Mitigation of Volcano Hazards, Springer, Berlin. pp 719-769
136. Watters RJ, Zimbelman DR, Bowman SD, Crowley JK (2000) Rock mass strength assessment and significance to edifice stability, Mount Rainier and Mount Hood, Cascade Range volcanoes. Pure Appl Geophys 157:957-976
137. Waythomas CF, Pierson TC, Major JJ, Scott WE (2013) Voluminous ice-rich lahars generated during the 2009 eruption of Redoubt Volcano, Alaska. J Volcanol Geotherm Res 259:389-413
138. Willingham WF (2005) The Army Corps of Engineers' short-term response to the eruption of Mount St. Helens. Oregon Hist Quart 106:174-203
139. Wisner B, Blaikie P, Cannon T, Davis I (2004) At Risk—Natural Hazards, People's Vulnerability and Disasters. Routledge, New York.

140. Witham C (2005) Volcanic disasters and incidents—a new database. J Volcanol Geotherm Res 148:191-233
141. Wood NJ, Soulard CE (2009) Community exposure to lahar hazards from Mount Rainier, Washington.
142. Wood N, Soulard C (2009) Variations in population exposure and sensitivity to lahar hazards from Mount Rainier, Washington. J Volcanol Geotherm Res 188:367-378

Chapter 4

Measuring Household Vulnerability to Climate-Induced Stresses in Pastoral Rangelands of Kenya: Implications for Resilience Programming

David C. Hall and Quynh Ba Le

Francis EO Opiyo[1,2], Oliver V Wasonga[1,3], and Moses M Nyangito[1]

[1]Department of Land Resources Management and Agricultural Technology, University of Nairobi, Kabete, Nairobi 00625, Kenya

[2]Dryland Development Centre, United Nations Development Programme (UNDP), United Nations Office in Nairobi, Nairobi 00100, Gigiri, Kenya

[3]German Institute for Tropical and Subtropical Agriculture (DITSL), Steinstraße 19, Witzenhausen 37213, Germany

ABSTRACT

This study uses statistical and econometric tools to measure households' vulnerability in pastoral rangelands of Kenya. It considered 27 socio-economic and biophysical indicators obtained from 302 households' in-depth interviews to reflect climate vulnerability components: adaptive capacity, exposure and sensitivity. The theoretical framework used combines exposure and sensitivity to produce potential impact, which was then compared with adaptive capacity in order to generate an overall measure of vulnerability. Principal component analysis (PCA) was used to develop weights for different indicators and produce a household vulnerability index (HVI) so as to classify households according to their level of vulnerability. In order to understand the determinants of vulnerability to climate-induced stresses, an ordered probit model was employed with predictor variables. The results show that 27% of households were highly vulnerable, 44% were moderately vulnerable and 29% of households were less vulnerable to climate-induced stresses. Factor estimates of the probit model further revealed that the main determinants of pastoral vulnerability are sex of household head, age of household head, number of dependents, marital status, social linkages, access to extension services and early warning information, complementary source of income, herd size and diversity, herd structure, herd mobility, distance to markets, employment status, coping strategies and access to credit. Therefore, policies that address these determinants of vulnerability with emphasis on women's empowerment, education and income diversifications are likely to enhance resilience of pastoral households.

BACKGROUND

Vulnerability, commonly defined as the propensity or predisposition to be adversely affected, has been studied as a composite of adaptive capacity, sensitivity and exposure to hazards (Adger and Kelly [1999]; Kelly and Adger [2000]; McCarthy et al. [2001]; Intergovernmental Panel on Climate Change IPCC [2001]; Adger [2006]; Füssel [2007]; Paavola [2008]; Yuga et al. [2010]). Adaptive capacity is the ability of people to cope with or adjust to the changing context and is explained by socio-economic indicators. Sensitivity is the ability of a system to be affected, and exposure is the incidences of events (Kasperson et al. [1995]; Adger [2006]; Paavola [2008]). Vulnerability is thus comprised of risks or a chain of risky events that households confront in pursuit of their livelihoods, the sensitivity of livelihood to these risks, the response or options that households have for managing these risks and finally the outcomes that describe the loss in wellbeing (Turner et al. [2003]).

On the other hand, resilience is seen as the ability to self-organize, learn and adapt to risk hazards (Carpenter et al. [2001]; Turner et al. [2003]). The United Nations International Strategy for Disaster Reduction (United Nations International Strategy for Disaster Reduction-UNISDR [2009]) defines resilience as the transformative process of a household or community exposed to hazards to resist, absorb, accommodate and recover from hazards in a timely and efficient manner. Turner et al. ([2003]) and Gallopın ([2006]) define hazards as threats to a system and comprised of stressors. For simplicity, the term stress is used in this study to denote a continuous or slowly increasing pressure (in this case drought), commonly within the range of normal variability. However, over the last decades, most of the scientific literature and discourses on vulnerability have concentrated on contributing to theoretical insights or measurements at the regional or national scale, with selected indicators for each region, and identifying resilience-building strategies that have implication for national and regional planning (Brooks et al. [2005]; Füssel [2007]; Hinkel [2011]). Yet, micro-level vulnerability analysis is an essential prerequisite for local-level planning and prioritization of resilience planning and strategies especially among the natural resource-dependent communities at risk to projected climate variability and changes (Callaway [2004]; Fraser et al. [2011]).While there is no superior scale of climate vulnerability analysis, recent studies by Yuga et al. ([2010]) and Marshall et al. ([2014]) have confirmed that micro-level analyses have hitherto largely been overlooked in favour of ecosystem-scale studies of biophysical vulnerability.

As observed by Deressa et al. ([2008]), climate change vulnerability analysis ranges from local or household level (Adger [1999]) to the global level (Brooks et al. [2005]; Intergovernmental Panel on Climate Change IPCC [2014]). However, the choice of scale is dictated by the objectives, methodologies and data availability. This present study focused on household-level vulnerability analysis. After all, it is by understanding, planning for and adapting to a changing climate that individual households can take advantage of opportunities to reduce risks associated with climate-induced stresses (Madu [2012]). Similarly, Klein et al. ([2007]) observed that vulnerability analysis to climate change is needed at the level that would enable policy makers to tackle climate change challenges with the precision that is necessary, particularly in the arid and semi-arid regions of Africa. Previous studies by Brooks et al. ([2005]) and Intergovernmental Panel on Climate Change IPCC ([2012]) concluded that the majority of households are particularly vulnerable to the impacts of climate change in the arid and semi-arid regions. Although the causes of such vulnerability are multi-dimensional, they are primarily due to widespread poverty, food insecurity, recurrent droughts, land degradation, inequitable land distribution and overdependence on rain-fed agriculture (Notenbaert et al. [2013]; Lo´pez-Carr et al. [2014]). Some of the other common factors also postulated to determine adaptive capacity, and therefore influence vulnerability at the household level, include access to resources, markets and infrastructure; household structure, gender, education and age; farm size; income and income diversity; access to community-based organizations, information, credit facilities, savings and loans; and health status among others. Stringer et al. ([2009]) argued that though all households in a community are exposed to risks associated with climate change and could potentially be rendered vulnerable, the poorer households are the most at risk of adverse impacts of climate variability and change. That notwithstanding, the proportion of households vulnerable to extreme weather events is perceived to be increasing in arid environments of eastern Africa, especially in Kenya, and there is an uncertain degree to which the population or the system is becoming susceptible and unable to cope with hazards and stresses, including the effects of climate change.

In Kenya, there is a consensus that projected climate change will worsen food security, mainly through increased extremes and temporal or spatial shifts (Eriksen and O'Brien [2007]; Herrero et al.[2010]; Sherwood [2013]). In fact, there is considerable potential impact of these global drivers of change on the production systems and resource-poor households who depend on them. Studies by Kabubo-Mariara ([2009]) and Silvestri et al. ([2012]) revealed that extreme weather events such as prolonged dry spells and intense rainfall are already affecting rural communities in parts of arid and semi-arid lands (ASALs) of Kenya. At present, nearly 30% of the total human population resides in the

ASALs, which cover approximately 88% of the country's land mass, and hold almost 70% of the total national livestock herd. However, large proportions of pastoralists who reside in ASALs are believed to be at risk of food production deficit, with potential declines in pasture and water availability (Opiyo et al. [2011]), all exacerbated by extreme climate events. Hence, the purpose of this paper is to identify the determinants of vulnerability and measure micro-level vulnerability of pastoralists in arid rangelands of Kenya. The vital information that is obtained from household-level vulnerability analysis is presently lacking in national and regional level assessments. Deressa et al. ([2008]), Pearson et al. ([2008]) and Sherwood ([2013]) show that vulnerability contexts are diverse for different multiple spatial scales, and therefore this approach can contextualize how climate variability and change affect pastoralist livelihoods.

This study was thus carried out to investigate households' vulnerability to climate variability and change to climate-induced stresses in a case study of a pastoralist rangeland of Kenya. The study identifies some of the determining factors for vulnerability based on certain household social, economic and environmental (biophysical) characteristics. The household was selected as the main unit of analysis because major decisions about adaptation to climate-induced stresses and livelihood processes are taken at that level (Thomas [2008]). Nevertheless, households are connected to the wider community, which can greatly influence their decision-making processes in relation to use of particular productive resources. The findings of this study can therefore be useful for targeting interventions, priority setting and resource allocations at micro-level. Complemented with studies analysing climate change impacts and findings from country-level adaptive capacity, governmental policy can be informed. At the same time, the uncertainty associated with climate variability and change demands an approach that prepares rural households without relying on detailed climate projections. The study therefore focused on the adaptive capacity of households for wider resilience programming. Moreover, the findings would assist in resources allocation and determination of resilience investment opportunities that are likely to increase the adaptive capacity of the vulnerable households. The findings in particular suggest potential avenues for research that may further enhance understanding of household vulnerability to climate stresses in the arid rangelands of Kenya.

Study Area

The study was conducted in Turkana County, northwestern rangelands of Kenya (Figure 1). The county lies between longitude 34° 30′ and 36° 40′E and latitude 1° 30′ and 5° 30′ N, covering an area of 77,000 km^2 with an estimated human population of 855,399 according to the 2009 Kenya housing and population census report. The county is one of the poorest in Kenya with an estimated 87.5% of the population living below the absolute poverty line (Kenya National Bureau of Statistics KNBS [2013]). In Lokichoggio, Kakuma and Oropoi divisions where the household survey was undertaken, poverty is exacerbated by the harsh environment, poor infrastructure and low access to basic services, in addition to other underlying causes of poverty that are experienced in northern Kenya. About 96% of the study area falls into arid and very arid eco-climatic zones (Jaetzold and Schmidt [1983]). In the study area, because of aridity, managing short-term climatic fluctuations as well as adapting to long-term changes is critical in sustaining livelihoods. According to the most recent household economy approach assessments, pastoralism is the predominant livelihood accounting for over 55% of the population; 15.6% agro-pastoral, 8% fisher folks and 8.4% are in the urban/peri-urban areas (Household Economy Approach HEA [2012]).

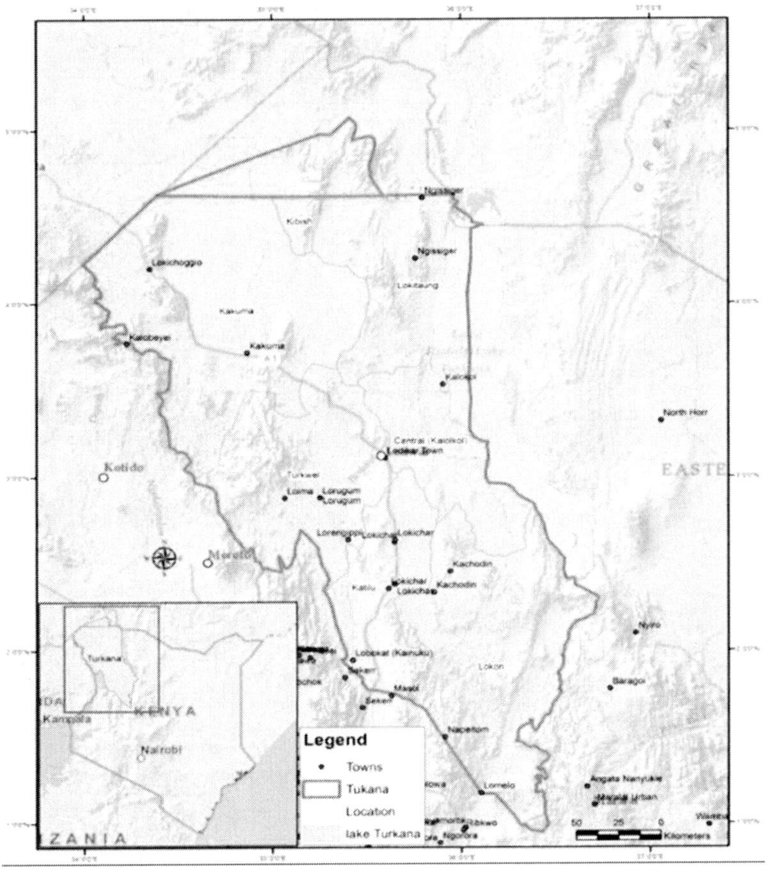

Figure 1: Location of the study area.

Rainfall in the study area exhibits a bimodal pattern, and distribution is unreliable and erratic in both space and time, with an average annual rainfall of 120 mm in the east to over 200 mm in the northwest. Figure 2 shows the annual rainfall variability of the study area. The main rainy season at local level between March 1950 and July 2012 the study area received very low and variable rain. In the past decades, major widespread drought events were experienced in 1980 to 1984, 1990 to 1995, 1999 to 2000, 2008 to 2009 and 2010 to 2011 within the study area. The last drought crisis in the Horn of Africa is estimated to have affected over 13 million people, including 3.75 million Kenyan, especially in the ASALs (UN-OCHA [2011]). The daily temperature ranges are extremely high with an average daily range varying between 24°C and 38°C with a mean of 30°C. The vegetation in this area is mainly scattered *Acacia* bush and a cover of annual herbaceous plants. The density of the woody plants such as *Acacia reficiens* and *A. mellifera* increases on hilly ground.

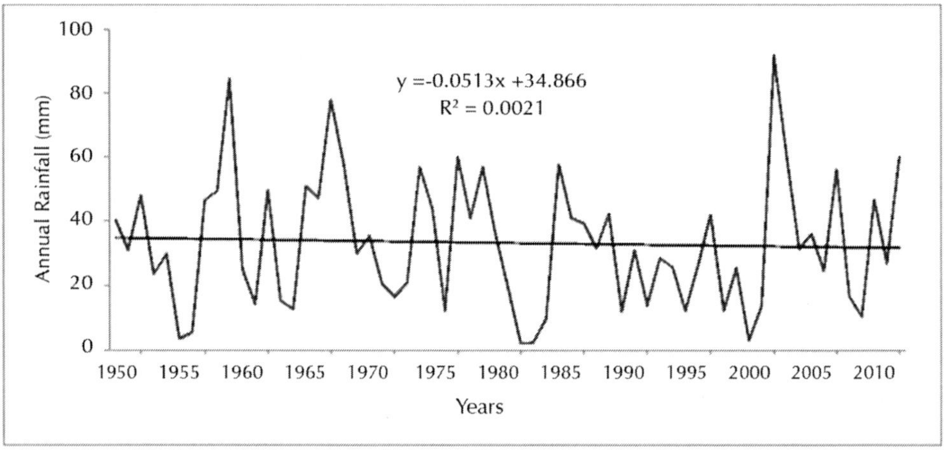

Figure 2: Annual rainfall variability for period 1950 to 2012 in the study area.

The study area is inhabited by the Turkana pastoralists whose major occupation is extensive livestock production (Watson and van Binsbergen [2008]).The livestock species kept include camels, cattle, sheep, goats and donkeys, which have different forage and water requirements and variable levels of resilience during drought episodes. In Turkana, droughts, famine, epidemics and flash floods are also very common occurrences. In most instances, these shocks and stresses are associated with climatic variability and change (Notenbaert et al. [2013]). The outcome of these climatic stresses has been high livestock mortality, loss of property, displacement of people and resources conflicts (Schilling et al. [2012]). In general, by weakening the productivity and functioning of livelihood resources, climatic stresses aggravate the vulnerability of households dependent on these resources. However, Turkana pastoralists are known to be highly mobile with no fixed residence (McCabe et al. [1988]), but their regular patterns of movement often get disrupted by conflicts and disease outbreaks (Hendrickson et al. [1998]; Schilling et al. [2012]).

METHODS

Data Collection

A total of 302 pastoralist households were selected for interviews a using semi-structured questionnaire to elicit responses between June and July 2012. Systematic purposive sampling was used to select drought-prone divisions in Turkana County which include Lokichoggio, Kakuma and Oropoi. However, the specific kraals (homesteads)/ locations where households were selected based on accessibility of the area. Consequently, 10 locations/kraals were selected from which the sample households were selected randomly proportional to population size. Table 1 shows the sampled households from each of the 10 kraals/ locations. A semi-structured questionnaire was used to interview the household heads. The questionnaire was divided into the following: demographic and economic household characteristics, livestock and crops production, access to extension services, credit access, hazards occurrence, perception level, adaptation strategies pursued, different coping strategies, level of resilience and other relevant information.

Table 1: Sampled households by Kraal

Location/Kraals	Number of households	Sampled household	**Percentage of respondent**
Mogila	1,536	20	6.62
Lokichoggio	1,868	9	2.98
Lopusiki	649	1	0.33
Lopwarin	277	11	3.64
Songot	459	19	6.29
Nanam	748	38	12.58
Kakuma	12,767	33	10.93
Letea	4,957	56	18.54
Kalobeyei	2,577	39	12.91
Pelekech	1,863	76	25.17

Kraal used herein to refer to a traditional pastoralists village of huts, typically enclosed by a fence. (Source: Authors' compilation).

Opiyo *et al.*

Opiyo *et al. Pastoralism* 2014 4:10, doi: 10.1186/s13570-014-0010-9

In addition, *in situ* rainfall and temperature datasets relevant for this study were obtained from Kenya Meteorological Services, National Drought Management Authority (National Drought Management Authority - NDMA [2013]) and County government officers in Lodwar, Turkana to strengthen some of the study findings. The STATA software was used to analyse the data obtained.

Analysis of Households' Vulnerability

There are various ways of analysing vulnerability, namely, socio-economic, biophysical and an integrated approach, which unites both socio-economic and biophysical factors. The socio-economic vulnerability assessment approach focuses on the socio-economic and political status of individuals or groups. Individuals in a community vary in terms of education, gender, wealth, health status, access to credit, access to information and technology, formal and informal (social) capital and political power, which are responsible for variations in vulnerability levels (Füssel [2007]; Deressa et al.[2008]). Consequently, vulnerability is considered to be a starting point or a state that exists within a system before it encounters a hazard event (Kelly and Adger [2000]). In this regard, vulnerability is shaped by society as a result of institutional and economic changes. The socio-economic approach focuses on identifying the adaptive capacity of individuals or communities based on their internal characteristics. One major limitation of this approach is that it focuses only on variations within society, but in reality, societies vary not only due to socio-political factors but also because of environmental or biophysical factors. The socio-economic approach does not account for the natural resource bases which have the potential to counteract the negative impacts of environmental shocks. For example, areas with easily accessible underground water can better cope with drought by utilizing this resource, compared to areas without it (Deressa et al. [2008]).

The second commonly used approach is biophysical that attempts to assess the level of damage that a given environmental stress causes on both social and biological systems. It is sometimes known as an impact assessment. The emphasis is on the vulnerability or degradation of biophysical conditions (Liverman [1990]). This is a dominant approach employed in studies of vulnerability to natural hazards and climate variability and change (Hewitt [1995]). Füssel ([2007]) identified this approach as a risk-hazard approach. The biophysical approach, although very informative, also has a major limitation that assessment of biophysical factors is not a sufficient condition for understanding the complex dynamics of vulnerability. This approach also neglects both structural factors and human agency in producing vulnerability and in coping or adapting to it. The approach overemphasizes extreme events while neglecting root causes and everyday social processes that influence differential vulnerability (Liverman [1990]; Hewitt [1995]; Pulwarty and Riebsame [1997]). The third approach is the integrated vulnerability analysis, which combines both the socio-economic and biophysical factors. This approach includes all the internal state of vulnerability and the external situation. This analytical approach was applied by Madu ([2012]) in agro-ecological based household vulnerability analysis in Ethiopia and by Deressa et al. ([2008]) in regionally based vulnerability analysis.

This present study replicates an integrated vulnerability approach to develop vulnerability indices for each household as proposed by Madu ([2012]) and adopted by Tesso et al. ([2012]) in Ethiopia. In this research, it is assumed that households with higher adaptive capacity are less sensitive to impacts of climate-induced stresses, thus keeping the level of exposure constant. The integrated assessment approach combines both socio-economic and biophysical approaches to determine vulnerability. The vulnerability index development is given as developed by the Intergovernmental Panel on Climate Change IPCC ([2012]) that vulnerability is seen as the net effect of adaptive capacity (socio-economic) and sensitivity/exposure (biophysical):

$$\text{Vulnerability} = (\text{Adaptive capacity}) - (\text{Sensitivity} + \text{Exposure}) \quad (1)$$

When the adaptive capacity of the household exceeds that of its sensitivity and exposure, the household becomes less vulnerable to climate change impacts and the reverse is also true. As explained in the foregoing, each set (adaptive capacity, sensitivity and exposure) is composed of different variables. The model specification further looks like

$$V_i = (A_1X_{1J} + A_2X_{2J} + \ldots + A_{2n}X_{nJ}) - (A_{n+1}Y_{1J} + A_{n+2}Y_{2J} + \ldots + A_{n+n}X_{nJ}) \quad (2)$$

where V_i is the vulnerability index, while X_S are elements of adaptive capacity, and Y_s are elements of exposure and sensitivity. The values of X and Y are obtained by normalization using their mean and standard errors. For instance, $X_{IJ} = (X_{IJ} - X_{1*}) / S^{1*}$, here X_{1j*} is the mean of X_{1j} across the different households, S_1* is its standard deviation. X_1 is the principal component result of factors. In this regard, the first principal component of a set of variables is the linear index of all the variables that captures the largest amount of information common to all the variables. The whole matrix of variables of adaptive capacity (X_{ij}) and variables of exposure and sensitivity (Y_{ij}) appears as follows:

$$X_{ij} / Y_{ij} = \begin{pmatrix} (X_{11} + X_{12} + \ldots + X_{2n}) - (Y_{11} + Y_{12} + \ldots Y_{2n}) \\ . \\ (X_{m1} + X_{m2} + \ldots + X_{mn}) - (Y_{m1} + Y_{m2} + \ldots Y_{mn}) \end{pmatrix} \quad (3)$$

The i and j in the foregoing notation imply the number of rows (in this case is the 302 individual households) and the number of columns (27 variables of adaptive capacity, exposure and sensitivity). In Equation 4, the A_S is the first component score of each variable computed using principal component analysis (PCA) in STATA. Finally, the vulnerability index of each household is obtained using Equation 4 as follows:

$$V_i = \begin{pmatrix} A_1 \\ A_{12} \\ . \\ A_{n+n} \end{pmatrix} \times \begin{pmatrix} (X_{11} + \ldots + X_{2n}) - (Y_{11} + \ldots + Y_{2n}) \\ . \\ (X_{m1} + \ldots + X_{mn}) - (Y_{m1} + \ldots + Y_{mn}) \end{pmatrix}_t \quad (4)$$

In calculating the direction of relationship in vulnerability indicators (that is, their sign), a negative value was assigned to both exposure and sensitivity. The justification is that households which are highly exposed to climate shocks are more sensitive to damage, assuming constant adaptive capacity. The implication is that a higher net value indicates lesser vulnerability and vice versa. However, in creating the indices, the scale of analysis is important. As noted by Tesso et al. ([2012]) from Deressa et al. ([2008]), vulnerability analysis ranges from local or household level to the global level (Brooks et al. [2005]). The choice of scale is dictated by the objectives, methodologies and data availability. In this study, the households were classified into three categories based on the value of their vulnerability index, which puts households into highly vulnerable, vulnerable and less vulnerable categories. However, the index computed is not based on the thresholds or presents an absolute value. It is a relative measure, representing the households' own perception of how they have been coping in the past compared to other households.

Factors Influencing Household Vulnerability

Notenbaert et al. ([2013]) note that many factors contribute to vulnerability, and these factors undermine capacity for self-protection, block or diminish access to social protection, delay recovery or expose some households to greater or more frequent hazards than other households. The analysis was performed using ordinal logistic regression analysis. The ordinal logit model is used when the outcome variable is categorized in an ordinal scale, as in this case where vulnerability is ordered as (1) highly vulnerable, which implies households for whom the difference between adaptive capacity and sensitivity/exposure is significantly negative; (2) moderately vulnerable, which means that households for whom the difference between adaptive capacity and sensitivity/exposure is nearly zero; and (3) less vulnerable, which means that the difference between adaptive capacity and exposure/sensitivity is significantly positive. In this study, sensitivity of households to climate-induced stresses is represented by its associated impacts, i.e., shortage of food, loss of water and pasture resources and conflicts faced by those households. In the case of exposure, since all households are assumed to be located in the same environment, exposure is almost uniform across the respondents residing in the study area.

This model is particularly useful in that it can show movement between vulnerability groups, explaining who moves in and out of vulnerability. Following Greene ([1997]), the reduced form of the ordinal logit model is given as

$$Y_j^* = X_j^l \beta + U_{lj} \quad (5)$$

where Y is the level of vulnerability and involves ordered outcome, that is, $Y=1$ was given to households that have a high level of vulnerability as observed by the negative value of adaptive capacity minus sensitivity/exposure; $Y=2$ was given to households having adaptive capacity nearly equal to their sensitivity/exposure; and $Y=3$ was given to households having their adaptive capacity exceeding their sensitivity and exposure. Y^* is the given state of vulnerability. The X_{ij} are the explanatory variables determining vulnerability level. The independent variables included in the model were sex and age of the household head, experience in the study area, household size and education level of the household head, dependency, marital status, social linkages and visits by extension officers, access to early warning information, non-farm income, herd size, herd structure, access to markets, property regimes, access to remittances, employment, coping strategies, herd diversity, credit access, herd mobility, climate change, experience in increased temperature, drought, floods, and wind, and natural hazards encountered in 5 years. B_S are parameters estimated and U_{ij} is the disturbance term. Y^* is unobserved, but what was observed in this study is

$$Y = 1 \text{ if } Y^* \leq \mu 2 \quad (6)$$

$$Y = 2 \text{ if } \mu 2 < Y^* \leq \mu 3 \quad (7)$$

$$Y = 3 \text{ if } \mu 3 < Y^* \quad (8)$$

Given the cumulative normal function $\Phi\ (\beta' x)$, the probabilities can be shown, thus,

Prob [$y=1$ or highly vulnerable] $= \Phi\ (-\beta' x)$,

Prob [$y=2$ or neutral level of vulnerability] $= \Phi(\mu 2 - \beta' x) - \Phi\ (\mu 3 - \beta' x)$

Prob [$y=3$ or less vulnerable] $= 1 - \Phi\ (\mu 3 - \beta' x)$

RESULTS

Hazards Reported by Households

The results show that 49% of households experienced drought over the past three decades, with 27.1% of the respondents reporting livestock disease outbreak, 19% mentioning cross-border inter-community conflicts and 4% citing riverine flash floods as the main hazard (Table 2). However, it was difficult for the respondents to differentiate between threats and hazards, which required clarification of the differences. Drought events were reported to be frequent hazards in the area and had devastating impacts on household livelihoods, pasture and water, which escalates the area's chronic conflicts, insecurity and food insecurity.

Table 2: Hazards identified by respondents

Hazard	Description of hazards	Number of respondents	Percentage of respondents
Floods	Flash floods along ephemeral rivers	13	4.3
Droughts	Widespread frequent droughts	148	49.0
Disease outbreak	Livestock diseases outbreak are common	82	27.1
Others i.e. conflicts, fire	Inter-community conflicts along the borders over resources	59	19.5

United Nations ([2004]) defines a 'hazard' broadly as "a potentially damaging physical event, phenomenon or human activity that may cause the loss of life or injury, property damage, social and economic disruption or environmental degradation". (Source: Authors' compilation).

Opiyo *et al.*

Opiyo *et al. Pastoralism* 2014 4:10, doi: 10.1186/s13570-014-0010-9

Inter-community and ethnic conflict related to scarce grazing land and water resources was reported to have increased inter-tribal animosity, often resulting in armed violent conflicts, which are predatory in nature and much more destructive. Households stated that violent conflicts regularly lead to heavy losses of lives and livelihoods, undermining human and livestock population mobility, as well as development efforts. Study findings show that outbreaks of *peste des petit ruminants* (PPR) locally referred to as *lomoo* is considered as the major hazard, having the highest impact on small ruminants compared to other endemic livestock diseases in the area. On further probing with community animal health workers and veterinary officers, based on the symptoms described by the pastoralist respondents, other diseases were identified as *etome* (mange), *emadang'* (worms infestation), *lukoi* (contagious caprine pleuropneumonia), *lomeri* (lump skin disease), *lokichum* (heart water), *etune* (sheep/goat pox), *lotorebwo or lokipi* (trypanosomiasis), *logooroi* (hemorrhagic septicaemia) and *loukoi* (contagious bovine pleuropneumonia).

Socio-Economic and Biophysical Vulnerability

The social and economic variables contributing to vulnerability are summarized in Table 3. Findings show that more than 80% of the respondents had no basic primary education, while the majority are not able to read and write. This in turn reduces a household's ability to understand climatic information, access market and early warning information. Data on household size showed that 64% of those interviewed had more than five persons, with 43% respondents reported more than five dependents. Whereas the sex of the household head is likely to be critical for climate adaptation, the study revealed that more than 50% of households are headed by females. In addition, 54% of the respondents indicated that they do not have access to livestock extension services. The results imply that the vulnerability level of households to the frequently occurring climate-induced stresses is largely determined by gender and education level of the household head in the study area.

Table 3: Social, economic and environmental indicators and their effect on vulnerability level

Hypothesized variables	Percentage	Influence on vulnerability[a]
Social vulnerability variables		
Sex of HH head: female-headed	50.7	+

Age of HH head: 50+ years	26.6	−
Experiences in the area: less than 5 years	7.9	+
HH size: more than 5 persons	64.6	−
Educational level: no primary education	80.1	+
Dependents: more than 5 persons	43.8	−
Marital status: single (including divorced and widowed)	29.5	+
Linkages: having no social linkages	25.8	−
Visit by extension officers: no access to extension services	54.3	+
Access to early warning information: no access to the information	74.8	+
Economic vulnerability variables		
Non-farm income: have no non-farm income	35.5	−
Herd size: own less than 2 TLUs	18.3	+
Herd structure: no milking herd	72.8	+
Distance to market: more than 10 km away	39.4	−
Property regime: own private land	8.6	−
Access to remittances: no cash transfers	58	+
HH employed: no member of HH employed	88.1	+
HH coping strategies: more than 2 coping strategies	92.1	+
Livestock diversity: less than 2 livestock species	45.4	−
Credit access: having no access to credit at all	77.8	+

Mobility: able to move livestock freely	67.9	−
Environmental vulnerability variables		
Climate change: experiencing change	96.7	+
Temperature: experiencing increase	47.9	+
Drought: noticed increasing events	3.4	+
Flood: noticed change	4.1	+
Wind: noticed unusual change	11.6	+
HH facing more than 2 hazards in 5 years	72.5	−

[a]Positive sign indicates that the variable increases vulnerability, while negative sign means it reduces vulnerability. TLU, tropical livestock unit (1 TLU=250 kg). (Source: Estimated from June to July 2012 household survey interviews in Turkana; Authors' compilation).

Opiyo *et al.*

Opiyo *et al. Pastoralism* 2014 4:10, doi: 10.1186/s13570-014-0010-9

More than 35.5% of the households had no diversified income sources and heavily depend on livestock as their source of income. Results show that only a few of the households were practicing dry land crop farming alongside livestock keeping. For the livestock owned, 18% of respondents had less than two tropical livestock units (TLUs[a]), and 72% of households had no milking herd during the interview period. Further, the result revealed that about 68% of households practice seasonal herd mobility and flexible resource use in the study area. The results show that 88% of households had none of its member formally employed, and 92% of respondents use more than two coping strategies to survive in this harsh climatic condition, despite the low access to credit, access to remittances and long distances to markets. Other economic concern mentioned was the high formal unemployment opportunities, especially for the youths in Turkana.

[a]A TLU is the 250-kg live weight of any domestic herbivore (sheep or goat = 0.1 TLU; donkey = 0.4 TLU; cattle = 0.7 TLU; camel = 1.0 TLU).

Table 3 displays environmental, economic and biophysical factors hypothesized for climate-induced vulnerability. The results show that the sensitivity and exposure to climate variability in terms of those experiencing climate variability (96.7%) and people facing more than two hazards in 5 years (72.5%) have contributed negatively to the vulnerability level of households. Approximately 48% of the respondents noticed an increase in temperature. However, most of the environmental variables increase household's level of vulnerability to climate-induced stresses as indicated by the positive sign.

Measuring Household-Level Vulnerability

Table 4 presents the result of principal component analysis and its association with the social, economic and environmental variables. The factor scores (weights) of the first principle component analysis was positively associated with the majority of the indicators identified under adaptive capacity, exposure and sensitivity. Holding exposure and sensitivity constant, a negative index shows the household to have relatively lower adaptive capacity when compared to a household with a positive index value and vice versa.

Table 4: Factor score for the first principal component analysis

Factors	Factor Score
Social vulnerability variables	
Sex of HH head: female-headed	0.22917
Age of HH head: 50+ years	−0.25804
Experiences in the area: less than 5 years	−0.02906
HH size: more than 5 persons	−0.29837
Educational level: no primary education	0.039479
Dependents: more than 5 persons	−0.31599
Marital status: single (including divorced and widowed)	0.07224
Linkages: having no social linkages	−0.04414

Visit by extension officers: no access to extension services	0.189862
Access to early warning information: having no access	0.19573
Economic variables	
Non-farm income: HH with no farm income	−0.34130
Herd size in TLU: own less than 2 TLUs	0.293813
Herd structure: no milking herd	0.100562
Distance to markets: more than 10 km away	−0.15870
Property regime: own private land	−0.06728
Access to remittances: no cash transfers	0.047677
HH employed: no member of HH employed	0.095173
HH coping strategies: more than 2 coping strategies	0.101335
Livestock diversity: less than 2 livestock species	−0.42800
Credit access: having no access to credit at all	0.178153
Mobility: able to move livestock freely	−0.14718
Environmental variables	
Climate change: experiencing change	−0.02276
Temperature: experiencing increase	0.081161
Drought: noticed increasing events	0.164723
Flood: noticed change	0.039066
Wind: noticed unusual change	0.259975
HH facing more than 2 hazards in 5 years	−0.06667

Abbreviations: Households (HH), Tropical livestock unit (TLU).

Opiyo *et al.*

Opiyo *et al. Pastoralism* 2014 4:10, doi: 10.1186/s13570-014-0010-9

To compute the vulnerability index in Equation 2, indicators of adaptive capacity, which are positively associated with the first principal component analysis, and indicators of sensitivity and exposure, which are negatively associated with the principal component analysis, were used in this study. The variables considered in Equation 2 include sex of the household head, education level, marital status, access to extension services and early warning information, livestock ownership, herd structure, access to cash remittances, household employment status, coping strategies and access to credit. However, for the exposure and sensitivity, all the variables were considered in the analysis. This is because adaptive capacity is considered as positively contributing to the reduction of vulnerability, while exposure and sensitivity are negatively contributing to vulnerability reduction. The larger the factor score the more important is the variable and contributes more to the household's vulnerability.

The households were classified into three categories using the vulnerability index: less vulnerable are households that are in a vulnerable situation but can still cope; moderately vulnerable households are those that need urgent but temporary assistance in case of shock and stresses; and the highly vulnerable are those households that are almost at a point of no return. The result shows that the majority of households fall within the moderately vulnerable category, with 44% households having an index from −1.00 to 1.00. The less vulnerable households had an index of 1.1 to 3.0 and constitute 29%, while the highly vulnerable households had an index of −0.9 to −3.0 but are 27% of the total households sampled (Table 5 and Figure 3). Although there seems to be normal vulnerability index distribution in Figure 3, a keen look at the values showed a slight shift to the left. Statistically, the households' distribution is skewed toward the highly vulnerable in the illustration. In general, the results reveal high and moderate vulnerability levels of the pastoralist households in Turkana.

Table 5: Classification of community according to the range of their vulnerability index

Vulnerability category	**Household situation**	**Vulnerability index**	**Percentage of households**
Highly vulnerable	Emergency level HHs	−0.9 to −3.5	27
Moderately vulnerable	Needs urgent but temporary external assistance to recover	−1.0 to +1.0	44

Less vulnerable	In a vulnerable situation but still able to cope	+1.1 to +3.0	29
Total			100

(Source: Computed from household questionnaire interviews in June to July 2012).

Opiyo *et al.*

Opiyo *et al. Pastoralism* 2014 4:10, doi: 10.1186/s13570-014-0010-9

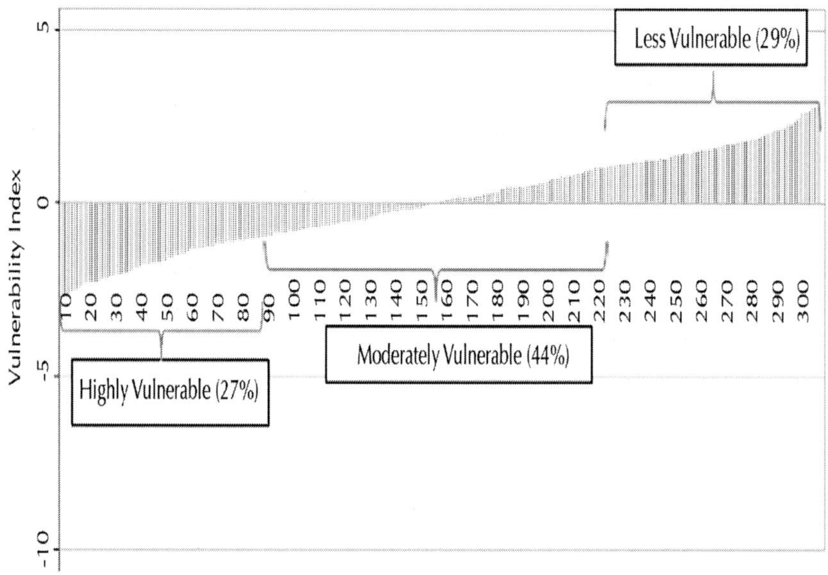

Figure 3: Vulnerability index by households in Kakuma, Oropoi and Lokichoggio divisions in Turkana.

Determinants of Vulnerability

The majority of highly vulnerable households were headed by females or someone with no primary level of education (Table 6). The highly vulnerable households also had more than five dependents, no access to early warning information, no milking herd and own less than two livestock species, and perceived changes in climate. By contrast, a household is likely to be less vulnerable when they are headed by a male, the household head is literate, not divorced or widowed, when they have access to extension services and early warning information, and own large and mixed-species herds. Similarly, households with access to cash remittances, with a member of the household in formal employment, and those with diverse coping strategies and access to credit facilities are reported to be less vulnerable to climate extremes. The results suggest that because of scarce resources during extreme climate events, households with high number of dependents and without any other livelihood diversification activities are likely to be more vulnerable than households with complementary sources of income and less dependents.

Table 6: Statistical description of model variables by vulnerability category

Vulnerability variables	Vulnerability category		
	Less vulnerable	Moderate vulnerable	Highly vulnerable
Sex of HH head: female	67 (44%)	62 (20%)	24 (45%)
Age of HH head: 50+ years	6 (41%)	33 (34%)	35 (25%)
Experiences in the area: less than 5 years	10 (16%)	10 (44%)	4 (4%)
HH size: more than 5 persons	27 (8%)	92 (22%)	76 (42%)
Educational level: no primary education	72 (45%)	107 (5%)	63 (55%)
Dependents: more than 5 persons	10 (47%)	58 (41%)	59 (33%)
Marital status: single (including divorced and widowed)	31 (42%)	41 (55%)	17 (46%)
Linkages: HH with no social linkages	18 (42%)	34 (52%)	26 (21%)
Visit by extension officers: no access to extension services	67 (17%)	65 (39%)	32 (20%)

Access to early warning information: having no access	77 (14%)	100 (9%)	49 (45%)
Non-farm income: HH with no farm income	5 (47%)	43 (33%)	58 (34%)
Herd size in TLU: own less than 2 TLUs	23 (39%)	17 (44%)	4 (31%)
Herd structure: no milking herd	72 (30%)	97 (23%)	51 (42%)
Distance to market: more than 10 km away	20 (44%)	51 (18%)	40 (27%)
Property regime: own private land	4 (26%)	13 (46%)	9 (40%)
Access to remittances: no cash transfers	58 (8%)	71 (36%)	46 (30%)
HH employed: no member of HH employed	81 (46%)	118 (15%)	67 (30%)
HH coping strategies: more than 2 coping strategies	82 (46%)	121 (50%)	66 (25%)
Livestock diversity: less than 2 species	5 (35%)	57 (35%)	75 (58%)
Credit access: having no access to credit at all	78 (46%)	108 (33%)	49 (17%)
Mobility: able to move livestock freely	42 (19%)	93 (41%)	70 (32%)
Climate change: experiencing change	43 (23%)	59 (26%)	38 (47%)
Drought: noticed increasing events	4 (44%)	3 (30%)	3 (21%)
Flood: noticed change	3 (33%)	7 (44%)	2 (29%)
Wind: noticed unusual change	11 (41%)	16 (25%)	7 (45%)
HH facing more than 2 hazards in 5 years	63 (40%)	99 (30%)	57 (26%)

HH, households; TLU, tropical livestock unit.

Opiyo *et al.*

Opiyo *et al. Pastoralism* 2014 4:10, doi: 10.1186/s13570-014-0010-9

As shown in Table 6, the majority of households in the moderately vulnerable category are the ones with less than 5 years' experience, either divorced or widowed household heads, household heads with no social linkages, household heads with no access to extension services, households who own less than two TLUs, own private lands, households which do not received any cash remittances, household heads with more than two coping strategies, households who practice mobility and are able to move freely with their livestock herd, and perceive climatic changes in the area.

The results of the ordered logistic regression model for all the single predictors' variables influencing a household's vulnerability are summarized in Table 7. A number of variables were statistically significant at 1% and 5% levels of significance and were important in influencing households' vulnerability to climate-induced stresses. The variables that showed significant influence on vulnerability include gender of the household head, age of the household head, size of the household, number of dependents, marital status, access to extension services and early warning information, complementary sources of income, herd diversity and structure, coping strategies and access to credit.

Table 7: Factors influencing household's vulnerability

Variables	**Estimate**	**SE**	**OR**	*z*	*pvalue*
Sex of HH head: female	−1.3828	0.2303	0.2509	−6.0049	<0.0001*
Age of HH head: 50+ years	1.5714	0.2704	4.8134	5.8106	<0.0001*
Experiences in the area: less than 5 years	−0.6187	0.399	0.5386	−1.5506	0.1220
HH size: more than 5 persons	2.2155	0.2651	9.1663	8.3585	<0.0001*
Educational level: no primary education	−0.1754	0.2684	0.8391	−0.6534	0.5140

Dependents: more than 5 persons	2.1434	0.2639	8.5281	8.1212	<0.0001*
Marital status: single (including divorced and widowed)	−0.4689	0.2349	0.6257	−1.9965	0.0468*
Linkages: having no social linkages	0.4232	0.2456	1.5268	1.7233	0.0859**
Visit by extension officers: no access to extension services	−1.0717	0.2237	0.3424	−4.7909	<0.0001*
Access to early warning information: having no access	−1.017	0.2535	0.3617	−4.0119	<0.0001*
Non-farm income: HH with no farm income	2.3332	0.2715	10.3108	8.5946	<0.0001*
Herd size in TLU: own less than 2 TLUs	−1.8785	0.3414	0.1528	−5.502	<0.0001*
Herd structure: no milking herd	−0.668	0.2434	0.5128	−2.7442	0.0064*
Distance to market: more than 10 km away	0.7694	0.2301	2.1586	3.3442	<0.0001*
Property regime: own private land	0.5666	0.375	1.7623	1.5111	0.1318
Access to remittances: no cash transfers	−0.2696	0.2161	0.7637	−1.2477	0.2131
HH employed: no member of HH employed	−0.6226	0.3337	0.5365	−1.866	0.0630**
HH coping strategies: less than 2 coping strategies	−1.0896	0.4268	0.3363	−2.5529	0.0112*
Livestock diversity: less than 2 livestock species	3.3875	0.3506	29.592	9.663	<0.0001*
Credit access: having no access to credit at all	−1.1825	0.2703	0.3065	−4.3753	<0.0001*
Mobility: able to move livestock freely	1.2984	0.2424	3.6635	5.3571	<0.0001*
Temperature: experiencing increase	−0.0412	0.2175	0.9597	−0.1892	0.8500
Drought: noticed increasing events	−0.2123	0.6269	0.8087	−0.3387	0.7350
Flood: noticed change	−0.1252	0.5225	0.8823	−0.2396	0.8108
Wind: noticed unusual change	−0.2462	0.3355	0.7818	−0.7338	0.4636

HH facing more than 2 hazards in 5 years	−0.0369	0.2409	0.9638	−0.1532	0.8784

SE, standard error; OR, odds ratio; *z*, score of two sample tests (basically a *t* test) though using the standard normal to calculate the *p*value and expressing the statistically significance levels at *5% and **10%. (Source: Ordinal logistic regression model: single predictors).

Opiyo *et al.*

Opiyo *et al. Pastoralism* 2014 4:10, doi: 10.1186/s13570-014-0010-9

The higher odds ratios computed for age of the household head, household size, dependents, complementary income sources, herd diversity and mobility imply that these factors have more effects on the households' vulnerability level. But it should be noted that majority of the sampled population derived their means of subsistence from nomadic pastoral production system which are weather sensitive and therefore absence of non-agricultural income sources (salary or wages labour) which are not weather sensitive would be a significant determinant on vulnerability level as reflected in the odds ratio figures. Herd diversity and mobility allow households to spread risk and pursue various coping strategies. Herd mobility was observed to be a significant (at 5%) determinant of households' vulnerability to climate variability and change.

From the ordered regression results, 75% of female-headed households are more likely to be vulnerable to climate-induced stresses and shocks compared to the male-headed households. This might be because female-headed households normally face gender discrimination with respect to resources, rights, education, income and economic opportunities. Household age was also an important demographic factor determining how vulnerable a household could be. For example, households headed by person above 50 years of age are more likely to be vulnerable compared with the younger persons. Consequently, elderly household heads are probably worse off in terms of preparing strategies to cushion their families against adverse climatic stresses and impacts and likely to make them more vulnerable. Similarly, the more dependents a household has the more likely for it to be vulnerable since a larger proportion of household resources are directed to dependents who cannot contribute much toward household welfare. Further results also show that widowed or divorced single-headed families are 37.4% more likely to be vulnerable than families where both spouses are present. According to the study results, it was observed that livestock diversity had higher odds ratio of 29.592, which suggest that households with more than two livestock species are 28 times more likely to shift from highly vulnerability category to moderately vulnerability level.

Although 27 variables were hypothesized to be correlated with vulnerability, ordinal regression model result confirmed that only 17 factors were significant (at $p<0.1$ and $p<0.05$) in influencing households' vulnerability. Although not computed in this present study, it should be noted that different determinants have differential effects on a household's vulnerability levels.

DISCUSSION

Drought was considered by the respondents as one of the most frequent hazards in Turkana, in addition to other hazards such as conflicts, disease outbreaks and flooding. The pastoralist respondents maintained that frequent, prolonged drought events have increased in severity over the past decade or so. This perception corroborate Nicholson's ([2014]) assertion about drought events in northern Kenya, with rainfall being at least 50% to 75% below normal in almost half of the drought-stricken region and is likely to be linked to changing climatic conditions within the greater Horn of Africa. However, there is a great deal of uncertainty and tension about when, where and how much climate change will happen. Even less is known about the actual impacts of these perceived changes (Kabubo-Mariara [2009]; Notenbaert et al. [2013]; Intergovernmental Panel on Climate Change IPCC[2014]) to the dry land communities. It is argued that due to the frequency of shocks in the study area, adaptive or coping capacity and mechanisms adopted by vulnerable households could well have equal or larger ranges to that of exposure and sensitivity. As discussed earlier on vulnerability, the interaction of environmental (biophysical) forces determines exposures and sensitivities, and various social, cultural, political and economic forces shape adaptive capacity (Yohe and Tol [2002]; Turner et al. [2003]; Skjeflo [2013]).

In the study area just like the rest of northern Kenya, pastoralist households largely bear the brunt of negative impacts of extreme climate events like drought, which include increased poverty, water scarcity, resource-based conflicts, disease outbreaks and food insecurity. The majority of people in Turkana live below the absolute poverty line (an estimated 87.5% of the population) with more than 50% heavily relying on food aid and safety net programmes from year to year. The people who are already poor in this remote part of the country are struggling to cope with the added burden of increasingly unpredictable weather, which is triggered by climate variability and change. Evidence shows that Turkana is constrained by the harsh climatic conditions and remoteness, coupled with poor infrastructure and low access to essential services (Republic of Kenya [2012]) Previous studies (Thornton et al. [2006]; Kabubo-Mariara [2009]) in the region confirmed that it is becoming increasingly difficult for households to bounce back from ever-changing, inconsistent weather affecting their livelihoods, and many have been forced to pursue other livelihoods and coping mechanisms that only increase the cycle of vulnerability.

This study has shown that female-headed households, households with experience of less than 5 years in the area, household heads with no primary level of education and households headed by divorced and widowed persons, with no access to extension services and early warning information, in particular, are disproportionately likely to be affected by climate stresses and variability. In times of climate stresses and shocks like drought, these categories of households tend to have fewer options to find other ways of making a living, because their very low levels of literacy reduce their opportunities in coping mechanisms such as wage employment. Similarly, female or divorced and widowed household heads are likely not to be empowered enough in pastoral communities to make household decisions (Nabikolo et al. [2012]) and are frequently without access to credit services and adequate capital assets or not able to own large herds to manage households' daily requirements. Similar observations have been made by Kakota et al. ([2011]) in Malawi and Tesso et al. ([2012]) in Ethiopia that widowed or divorced household heads are more vulnerable because they rely on income earned by either the father or mother as the bread winners. These findings make a strong case for continuous targeting of pastoralist women in resilience-building interventions in the rangelands.

The results are consistent with previous findings (Deressa et al. [2008]) in similar ecosystems. However, for the biophysical variables, the greater the level of household reliance on natural resources, such as pastoralism or dry land crop farming, the greater will be their vulnerability to climate variability and change. This is partly because the use of such natural resources is dependent on rainfall, which is projected to change. This study observed that almost all the postulated biophysical/environmental variables contribute positively to household vulnerability. It is likely that the level of dependence on natural resources especially pastures and water will vary from household to household. For example, while the majority of households (78.9%) depend on livestock herding as their main source of livelihood, for others, livestock is just an equal or lesser contributor besides other economic activities.

The level of vulnerability is closely associated with the degree of poverty in Turkana. This county has higher degrees of vulnerability in terms of households' adaptive capacity or social and economic characteristics that are linked to their exposure and sensitivity to biophysical variables (Lo´pez-Carr et al. [2014]). The findings of this study suggest that in case of an external stress or shock, the majority of the households would need some assistance for them to recover. These results reflect the findings of studies by Kenya National Bureau of Statistics (KNBS) ([2013]) which showed that Turkana is the poorest county in Kenya, with 87.5% of the population living in absolute poverty. Other studies have indicated that Turkana region has a number of households with high of dependents, low engagement in off-farm activities and low levels of education (Blench [2000]; McPeak and Barrett [2001]; Watson and van Binsbergen [2008]).

The determinants of households' vulnerability were found to be significantly influenced by the sex of the household head, age of the household head, size of the household, number of dependents, marital status, social linkages, access to extension services and early warning information. In addition, non-farm income, herd size and diversity, herd structure and herd mobility, access to markets, households' employment status, coping strategies and access to credit were also observed to be the key determinants of the households' vulnerability to climate-induced stresses. This concurs with studies by Eriksen et al. ([2005]) and Notenbaert et al. ([2013]) which similarly observed some of these factors to be the key determinant of households' vulnerability to climate variability and change in rural communities. The results are also consistent with previous findings by Kakota et al. ([2011]) and Gebrehiwot and van der Veen ([2013]). From these findings, it seems there is still more to be done to understand vulnerability and its underlying processes. In this paper, the causal relationships between the statistically significant variables and outcomes in terms of vulnerability levels are explained using existing literature. Some of these explanations are, however, rather speculative and not confirmed for the local situation. More in-depth qualitative fieldwork, such as including open-ended questions, conducting focus-group discussions or in-depth interviews with selected households based on these factors, could strengthen our confidence in some of these explanations. The combination of quantitative surveys with qualitative autopsy is likely to provide a much more in-depth analysis of determinants of vulnerability.

CONCLUSIONS

Understanding vulnerability to environmental change and extreme climate events is necessary for policy makers to develop mitigation and adaptation programmes for long-term resilience. Vulnerability analyses contribute to the knowledge on climate-sensitive socio-economic or ecological systems, enabling policy to be targeted on the most vulnerable places, sectors or people and adaptation options to be defined. The results of this study from Turkana portray social and biophysical vulnerability indices which are useful to local development programming for long-term resilience. However, as a prerequisite for building households' resilience to climatic extremes, in-depth understanding is necessary of the adaptive capacity, exposure and sensitivity.

This study focused on the micro-level to assess the variability of vulnerability across different households. Categorization of vulnerability levels was used to help identify households that are not vulnerable currently but have a high probability of becoming vulnerable in the future. In the future, studies assessing household-level vulnerability to climate-induced stresses should explore the use of panel data as well as cross-sectional data to portray longitudinal and cross-sectional characteristics of households. Similarly, future resilience interventions should target individual households within a community because major decisions about adaptation to climate-induced stresses and livelihood processes are taken at the micro-level.

We conclude that because of various social, economic and biophysical determinants observed to influence households' vulnerability to climate-induced stresses, policies with emphasis on women's empowerment, promoting education, supporting extension services and enhancing diversifications of income sources and access to credit, supporting herd mobility and diversity, creating employments, and increasing access to markets and early warning information are likely to improve resilience of pastoral households. Although the results of this study are specific to Turkana County in Kenya, the approach and findings could be applicable to other arid and semi-arid areas in the region.

AUTHORS' CONTRIBUTIONS

FEOO collected the data and drafted the manuscript. OVW and MN guided the design of the research study and provided technical reviews of the paper. All authors read and approved the final manuscript.

ACKNOWLEDGEMENTS

The work has been supported by small grants from the African Climate Change Fellowship Program, Global Change for System Analysis, Research & Training (ACCFP/START), and additional support from the Red Cross/Red Crescent Climate Centre/START and United Nations International Strategy for Disaster Reduction (UNISDR) under Climate and Development Knowledge Network small grant. The authors also thank Boniface Makone of the Welcome Trust for the statistical guidance during the preparation of this paper.

REFERENCES

1. Adger, WN, and PM Kelly. 1999. Social vulnerability to climate change and the architecture of entitlements. Mitigation and Adaptation Strategies for Global Change 4: 253–266.

2. Adger, WN. 1999. Social vulnerability to climate change and extremes in coastal Vietnam. World Development 27: 249–269.
3. Adger, WN. 2006. Vulnerability. Global Environmental Change 16 (3): 68–281.
4. Blench, R. 2000. Extensive pastoral livestock systems: Issues and options for the future. GCP/JPN/005/ JPN. FAO (Food and Agriculture Organization of the United Nations), Rome, Italy.
5. Brooks, N, WN Adger, and PM Kelly. 2005. The determinants of vulnerability and adaptive capacity at the national level and the implications for adaptation. Global Environmental Change 15 (2): 151–163.
6. Callaway, JM. 2004. Adaptation benefits and costs:how important are they in the global policy picture and how can we estimate them? Global Environmental Change 14: 273–284.
7. Carpenter, S, B Walker, JM Anderies, and N Abel. 2001. From metaphor to measurement: resilience of what to what? Ecosystems 4 (8): 765–781.
8. Deressa, T, RM Hassan, and C Ringler. 2008. Measuring Ethiopian farmers' vulnerability to climate change across regional states, IFPRI discussion paper no. 806.
9. Eriksen, S, and K O'Brien. 2007. Vulnerability, poverty and the need for sustainable adaptation measures. Climate Policy 7: 337–352.
10. Eriksen, S, K Brown, and K Mick. 2005. The dynamics of vulnerability:locating coping strategies in Kenya and Tanzania. Geographical Journal 171: 287–305.
11. Fraser, EDG, AJ Dougill, K Hubacek, CH Quinn, J Sendzimir, and M Termansen. 2011. Assessing vulnerability to climate change in dryland livelihood systems:Conceptual challenges and interdisciplinary solutions. Ecology and Society 16 (3): 14.
12. Füssel, H. 2007. Vulnerability:a generally applicable conceptual framework for CC research. Global Environmental Change 17: 155–167.
13. Gallopın, GC. 2006. Linkages between vulnerability, resilience, and adaptive capacity. Global Environmental Change 16: 293–303.

14. Gebrehiwot, T, and A van der Veen. 2013. Climate change vulnerability in Ethiopia:disaggregation of Tigray Region. Journal of Eastern African Studies 7 (4): 607–629.
15. Greene, WH. 1997. Econometric analysis. Prentice Hall, New Jersey.
16. Hendrickson, D, J Armon, and R Mearns. 1998. The changing nature of conflict and famine vulnerability: the case of livestock raiding in Turkana District, Kenya. Disasters 22: 185–199.
17. Herrero, M, C Ringler, J van de Steeg, P Thornton, T Zhu, E Bryan, A Omolo, J Koo, and A Notenbaert. 2010. Climate variability and climate change and their impacts on Kenya's agriculture sector. International Livestock Research Institute (ILRI), Nairobi, Kenya.
18. Hewitt, K. 1995. Excluded perspectives in the social construction of disaster. International Journal of Mass Emergencies and Disasters 13: 317–319.
19. Hinkel, J. 2011. Indicators of vulnerability and adaptive capacity: towards a clarification of the science–policy interface. Global Environmental Change 21: 198–208.
20. 2012. Livelihood profiles:Six livelihood zones in Turkana County, Kenya. June 2012 report prepared for Oxfam and Save the Children UK.
21. 2001. Climate change 2001:Impacts, adaptations and vulnerability. Report of working group II. Synthesis report. Cambridge University Press, Cambridge.
22. 2012. Managing the risks of extreme events and disasters to advance climate change adaptation. A special report of Working Groups I and II of the Intergovernmental Panel on Climate Change. Cambridge University Press, Cambridge, UK.
23. 2014. Climate change: Impacts, adaptation, and vulnerability. Working Group II contribution to the IPCC 5th assessment report.
24. Jaetzold, R, and H Schmidt. 1983. Farm management in Kenya, vol. II part C: East Kenya. Ministry of Agriculture, Nairobi, Kenya.
25. Kabubo-Mariara, J. 2009. Global warming and livestock husbandry in Kenya:impacts and adaptations. Ecological Economics 68: 1915–1924.

26. Kakota, T, D Nyariki, D Mkwambisi, and W Kogi-Makau. 2011. Gender vulnerability to climate variability and household food insecurity. Climate and Development 3 (4): 298–309.

27. Kasperson, J, R Kasperson, and B Turner. 1995. Regions at Risk: Comparisons of Threatened Environments. United Nations University Press, New York.

28. Kelly, PM, and WN Adger. 2000. Theory and practice in assessing vulnerability to climate change and facilitation adaptation. Climatic Change 47 (4): 925–1352.

29. 2013. Exploring Kenya's inequality: Pulling apart or pooling together. National Bureau of Statistics and Society for International Development (SID), Nairobi, Kenya.

30. Klein, RJ, S Eriksen, LO Naess, A Hammill, TM Tanner, C Robledo, and K O'Brien. 2007. Portfolio screening to support the mainstreaming of adaptation to climate change into development assistance, Tyndall Centre Working Paper No. 102.

31. Liverman, DM, et al.. 1990. Vulnerability to global environmental change. In Understanding global environmental change, ed. Kasperson RE, 27–44. Center for Technology, Environment, and Development, Clark University, Worcester.

32. Lo´pez-Carr, D, GP Narcisa, EA Juliann, MM Jankowska, C Funk, G Husak, and J Michaelsen. 2014. A spatial analysis of population dynamics and climate change in Africa: potential vulnerability hot spots emerge where precipitation declines and demographic pressures coincide. Population Environment 35: 323–339.

33. Madu, IA. 2012. Partial vulnerability of rural households to climate change in Nigeria:Implication for internal security. Working paper number 2; May 2012. The Robert S. Strauss Center for International Security and Law, University of Texas, USA.

34. Marshall, NA, CJ Stokes, NP Webb, PA Marshall, and AJ Lankester. 2014. Social vulnerability to climate change in primary producers: a typology approach. Agriculture, Ecosystems and Environment 186: 86–93.

35. McCabe, JT, R Dyson-Hudson, PW Leslie, PH Fry, N Dyson-Hudson, and J Wienpahl. 1988. Movement and migration as pastoral responses to limited and unpredictable resources. In Arid lands today and tomorrow, ed. Whitehead EE, Hutchinson

CF, Timmerman BN, Varady RG, 727–734. Proceedings of an International Research and Development Conference, Tucson, Arizona, USA.

36. 2001. Climate change 2001: Impacts, adaptation and vulnerability. Cambridge University Press, Cambridge.
37. McPeak, JG, and CB Barrett. 2001. Differential risk exposure and stochastic poverty traps among east African pastoralists. American Journal of Agricultural Economics 83 (3): 674–679.
38. Nabikolo, D, B Bashaasha, MN Mangheni, and JGM Majaliwa. 2012. Determinants of climate change adaptation among male and female headed farm households in eastern Uganda. African Crop Science Journal 20 (2): 203–212.
39. 2013. Turkana County drought monitoring and early warning bulletin - December 2013.
40. Nicholson, SE. 2014. A detailed look at the recent drought situation in the Greater Horn of Africa. Journal of Arid Environments 104: 71–79.
41. Notenbaert, A, SN Karanja, M Herrero, M Felisberto, and S Moyo. 2013. Derivation of a household-level vulnerability index for empirically testing measures of adaptive capacity and vulnerability. Regional Environmental Change 13: 459–470.
42. Opiyo, FEO, SM Mureithi, and RK Ngugi. 2011. The influence of water availability on pastoralist's resource use in Kitui and Mwingi districts of Kenya. Journal of Human Ecology 35 (1): 43–52.
43. Paavola, J. 2008. Livelihoods, vulnerability and adaptation to climate change in Morogoro, Tanzania. Environmental Science and Policy 11 (7): 642–654.
44. Pearson, L, R Nelson, S Crimp, and J Langridge. 2008. Climate change vulnerability assessment: Review of agricultural productivity. CSIRO climate adaptation flagship working paper 08.1.
45. Pulwarty, RS, and WE Riebsame. 1997. The political ecology of vulnerability to hurricane-related hazards. In Hurricanes: Climate and socioeconomic impacts, 185–214. Springer, Berlin, Heidelberg, New York.

46. 2012. National policy for sustainable development of Northern Kenya and other arid lands; Sessional Paper No. 8 of 2012, Republic of Kenya.
47. Schilling, J, FEO Opiyo, and J Scheffran. 2012. Raiding pastoral livelihoods: Motives and effects of violent conflict in north-western Kenya. Pastoralism: Research, Policy and Practice 2: 25.
48. Sherwood, A. 2013. Community adaptation to climate change: exploring drought and poverty traps in Gituamba location, Kenya. Journal of Natural Resources Policy Research 5 (2–3): 147–161.
49. Silvestri, S, E Bryan, C Ringler, M Herrero, and B Okoba. 2012. Climate change perception and adaptation of agro-pastoral communities in Kenya. Regional Environmental Change 12 (4): 791–802.
50. Skjeflo, S. 2013. Measuring household vulnerability to climate change—Why markets matter? Global Environmental Change 23 (6): 1694–1701.
51. Stringer, LC, MS Reed, AJ Dougill, and C Twyman. 2009. Local adaptations to climate change, drought and desertification: Insights to enhance policy in southern Africa. Environmental Science and Policy 12: 748–765.
52. Tesso, G, B Emana, and M Ketema. 2012. Analysis of vulnerability and resilience to climate change induced shocks in North Shewa, Ethiopia. Agricultural Sciences 3: 871–888.
53. Thomas, RJ. 2008. Opportunities to reduce the vulnerability of dryland farmers in Central and West Asia and North Africa to climate change. Agriculture, Ecosystems and Environment 126 (1–2): 36–45.
54. Thornton, PK, PG Jones, T Owiyo, RL Kruska, M Herrero, P Kristjanson, A Notenbaert, N Bekele, and A Omolo. 2006. Mapping Climate Vulnerability and Poverty in Africa. Report to the Department for International Development, International livestock research institute, Nairobi, Kenya.
55. Turner, BL, RE Kasperson, PA Matson, JJ McCarthy, RW Corell, L Christensen, N Eckley, JX Kasperson, A Luers, ML Martello, C Polsky, A Pulsipher, and A Schiller. 2003. A framework for vulnerability analysis in sustainability science. Proceedings of the National Academy of Sciences of the United States of America 100 (14): 8074–8079.

56. 2009. Global assessment report on disaster risk reduction.
57. 2004. Living with risk: A global review of disaster reduction initiatives. United Nations International Strategy for Disaster Reduction, Geneva, Switzerland.
58. 2011. Horn of Africa Crisis: Situation Report No. 13: United Nations Office for Co-ordination of Humanitarian Affairs.
59. Watson, DJ, and J van Binsbergen. 2008. Livestock market access and opportunities in Turkana, Kenya. ILRI Research Report 3, 106. ILRI (International Livestock Research Institute), Nairobi, Kenya.
60. Yohe, G, and RSJ Tol. 2002. Indicators for social and economic coping capacity—moving toward a working definition of adaptive capacity. Global Environmental Change 12: 25–40.
61. Yuga, NG, PG Shivakoti, and RP Sylvain. 2010. Household-level vulnerability to drought in hill agriculture of Nepal: Implications for adaptation planning. International Journal of Sustainable Development and World Ecology 17 (3): 225–230.

Chapter 5

Farm Types and Their Economic Characterization in Complex agro-Ecosystems for Informed Extension Intervention: Study from Coastal West Bengal, India

Rupak Goswami[1], Soumitra Chatterjee[2], and Binoy Prasad[1]

[1]Integrated Rural Development and Management Faculty Centre, Ramakrishna Mission Vivekananda University, Narendrapur, Kolkata-700103, INDIA

[2]Department of Agricultural Economics, Bidhan Chandra Krishi Viswavidyalaya, Mohanpur, West Bengal, Nadia-741252, INDIA

ABSTRACT

Identification and characterization of farming systems simplify huge diversity of farm types in complex agro-ecosystems, which is of critical importance for precise technological intervention and informed policy support. Multivariate statistical techniques like Principal Component Analysis (PCA) and Cluster Analysis (CA) may be used for a wide variety of situations associated with farm typology delineation. The present study conducted in coastal saline India demonstratively established the usefulness of such methodology in identification of predominant farm types and their characterization. Data collected from 144 farm households through questionnaire survey could identify four predominant farm types with differential income sources and resource-base. The methodological perspective employed in the study may be used as a decision support tool by extension agencies. On other hand, a differentiated, holistic and broad-based extension intervention with suitable institutional arrangement will be needed to address the need of these identified farm types. This will lead to a reduced transaction cost of the agricultural research and extension systems in diverse ecosystems in India and many similar situations in the developing countries.

BACKGROUND

Adoption of new technologies in agriculture is of central interest to both academicians and policy makers since this is directly related to the efficiency of an agricultural research and extension system (Bozeman [2000]). Higher adoption ensures higher return on investment in research and development and creates a cycle of production-led impact on economy and rural livelihoods. Unfortunately, there are numerous examples of technologies with great potential that have not been accepted by farmers, especially the smallholders of the developing countries. Quite often, these technologies do not fit well into heterogeneous smallholder systems, which need specific technological solutions. Such inherent variability often influences farmers' response to various technologies that aim at improving farm productivity and natural resource management (Lal et al. [2001]; Emtage and Suh [2005]). Unfortunately, both in agricultural and social sciences, complexity and diversity has been under-perceived and undervalued resulting in their neglect, under-estimation and exclusion from government statistics and policy framework (Chambers et al. [1989]). The archetypal Green Revolution technologies and 'transfer-of-technology' paradigm has also historically failed to cater to the needs of these diverse resource-poor agro-ecosystems in the developing countries (Pender and Hazell [2001]). Extension offered blanket recommendation for wide geographical areas and was largely used as a deterministic 'dart gun' (Roling [1988]), i.e. 'take the technology and transfer it to farmers'. The heterogeneity of the farming systems for which different technologies are needed, has been ruefully ignored.

Thus, study of farm typology is of practical interest for precise and effective technological interventions. Farm typology study recognizes that farmers are not a monolithic group and face differential constraints in their farming decisions depending on the resources available to them and their lifestyle (Soule [2001]). Ellis ([1993]) observes that small farmers are always and everywhere typified by internal variations along many lines. Although every farm and farmer is unique in nature, they can be clustered into roughly homogeneous groups. Developing a typology constitutes an essential step in any realistic evaluation of constraints and opportunities that farmers face and helps forwarding appropriate technological solutions, policy interventions (Ganpat and Bekele [2001], Timothy [1994]; Vanclay [2005]), and comprehensive environmental assessment (Andersen et al.[2009]). Moreover, typology studies are of paramount importance for understanding the factors that explain the adoption and/or rejection of new technologies (Kostrowicki [1977]; Mahapatra and Mitchell [2001]).

The heterogeneity of farming systems are created by a host of biophysical (e.g. climate, soil fertility, slope etc.) and socio-economic (e.g. preferences, prices, production objectives etc.) factors (Ojiem et al. [2006]). Researchers have examined factors such as farm resources such as cash and labor (Tittonell et al. [2007]), infrastructure such as agency and markets (Tittonell et al. [2006]), management practices (Tittonell et al. [2005]) and technological level (Gómez-Limón et al. [2007]). Few others have used string of factors together to explain the heterogeneity of farming systems (Ojiem et al. [2006]; Bidogeza et al. [2009]; Guto et al. [2010]; Tittonell et al. [2010]; Rowe et al.[2006]). The selection of factors that define farm typology varies greatly from study to study and may be governed by the purpose of research. For example, farm typologies were used to study appropriate fertilizer application (Tittonell et al. [2006]), resource use efficiency (Zingore [2007]; Tittonell et al. [2007]), water use efficiency (Senthilkumar et al. 2009), or overall classification of farm types (Bidogeza et al. [2009]), IPM (Leeson et al. [1999]) and may sometimes be crop specific in nature (Mwijage et al. [2009]; Reckling [2011]). Most of the farm typology study has focused on socio-economic and agro-ecological factors for classification of farms. Economic factors have been less used, especially in small-scale studies, for classifying farms (USDA, ERS [2000]; Briggeman et al.[2007]; Andersen [2009]). With economic characterization of farms as the objective at hand, the present study assumes that classification of farms based on economic returns from farm enterprises, along with other related non-economic factors, will give more effective insights into the farm type identification.

Indian farmers are heterogeneous in terms of agro-ecology and resource endowments and the transfer of appropriate technology requires careful targeting. This problem is exacerbated in complex-diverse-risk-prone coastal agro-ecosystems of the country covering approximately 10 Mha cultivated area, which is characterized by a host of bio-physical and socio-economic constraints coupled with diversity in livelihoods. At present, agricultural research is based on primarily 15 agro-climatic zones and 127 sub-regions based on comprehensive research review of states (Khanna [1989]). Socio-economic and structural factors are not reflected in such classifications. Economists classify farmers based on landholdings, which is not comprehensive enough for classifying complex farming systems. Indian Council of Agricultural Research (ICAR) used several factors for classification of farming systems, but the bases were more discussion-based with limited statistical applications, perhaps due to functional purpose of the classification (Sanjeev et al. [2010]).

At the micro level, classification of farmers might be of practical use for localized technological solutions and extension support. During the height of farming system research and extension paradigm, conceptualization of recommendation domain wanted to address this issue. A recommendation domain is a group of farmers whose circumstances are similar enough that they are eligible for the same recommendation (Harrington and Trip [1984]). This led to informed decisions in part of technology managers and higher rate of technology integration in smallholder systems. Classification of farming situation by the farmers has also been suggested by some authors (Conroy and Sutherland [2004]; Goswami et al. [2012]). However, these were more project-based and were hardly mainstreamed in the national research and extension systems in the developing countries (Frankenberger et al. [1989]), India being no exception. Participatory research is not mainstreamed in National Agricultural Research System in India and state-owned monolithic extension system is not prepared to deal with the need of small farms in diverse ecosystems (Glendenning et al. [2010]). This leads to poor adoption of technology and large yield gap (Aggarwal et al. [2008]). Extension support and policy intervention also suffers due to lack of informed decision by public extension. Although there has been experimentation with reorganized and decentralized systems of technology assessment and refinement and revitalized public extension systems, development of sound analytical tools for targeting extension has remained undermined till date. A sound methodology for profiling farm typologies will help in rapid transfer of appropriate technology, precise extension support and development of policy environment sensitive to immense diversity of smallholder farms in coastal saline India.

Macro-economic perspective on this systematic problem leads us to the examination of efficiency in agricultural research and extension system. Ruttan and Hayami ([1984]) cites the shortcomings of 'induced innovation' model (Hayami and Ruttan [1981]) that considers factor prices and user demand to induce the development of appropriate technology and the widely cited 'diffusion of innovations' theory (Rogers [1962], [1995]) that assumes diffusion to have taken place in an institutional vacuum. Unlike the neoclassical theory in economics, New Institutional Economics recognized the importance of institutions in explaining uneven performance of economies (North [1997]) and suggested that this type of analysis can lead to the necessary institutional developments vis-à-vis technical developments generated by R&D systems (Kydd [2002]). Within the extension science literature, responses could be traced in the form of Agricultural Knowledge and Information System (AKIS) concept (Engel [1995]) that sees agricultural innovation as a function of multi-stakeholder process. Roling ([1988]) departed from Rogers' theory towards a precise technological targeting and drew on the marketing research tradition (Kotler [1986]) for the same.

Increasing focus on the role of institutions in differential economic growth has now added new dimension to the study of technology transfer and appropriate technology (Hall and Yoganand[2004]; Radulovic [2005]). New Institutional Economics (NIE) that makes explicit considerations of institutional arrangement and institutional environment (North [1997]; Williamson [1990]), are now frequently drawn on in the literature on efficient extension mechanisms (that reduces transaction cost and enhances economic efficiency) (Birner and Anderson [2007]; Kherallah and Kirsten [2002]). However, the application of NIE was mostly focused on the budgetary constraints and institutional evolution, markets, and institutional development (Pal et al. [2003]). Methodology for identification of farm types for precise technological intervention and policy support, if mainstreamed, may be beneficial to the institutional arrangement that reduces transaction cost in smallholder agricultural production by providing appropriate technology to the smallholders rapidly. This is even more important in the regime of open economy where smallholders will have to be served with less transaction cost for long-term economic and environmental sustainability.

The present study was undertaken to identify the predominant farm types in coastal agro-ecosystem of India and to characterize them by some important socio-economic indicators. The article demonstrates the methodology of farm typology study when farming systems are heterogeneous and in need of appropriate technology for agricultural sustainability.

METHODS

Area of the Study

West Bengal is an eastern state of India and one of the major players in agricultural production. It has a high population density (976 per sq km) and varied farming systems. Agriculture is the mainstay of rural livelihoods and nearly 90% of the cultivators are small and marginal farmers jointly holding 84% of the State's agricultural lands. Another 3 million landless families have earned the right to cultivate and grow crops on their own land after enactment of Operation Barga system (land reform movement) by the State Government (GoWB [2005]). The high population density has resulted in unabated fragmentation of cultivable land and insufficient farm size to sustain livelihoods. Moreover, increase in the price of agricultural inputs, uncertain price of perishable agricultural produce, inadequate market infrastructure, distress sale of produce by small and marginal farmers are some of the problems that pose serious challenges to sustainability of farm sector in the state. The socio-economic condition of the farming community is declining in the absence of appropriate technical, social, financial and market interventions (NABCOM [2009]).

Based on climate, soil and physiography, there are six agro-climatic Zones for West Bengal. Among these, South 24 Parganas district comes under Coastal Saline Zone (Figure 1) (Gajbhiye and Mandal[2008]), one of the biggest Zones with high cropping intensity (143%) i.e. the ratio of gross and net cultivated area in percentage terms, and covering part or whole of 6 districts of West Bengal. Climate is tropical moist sub-humid with 1796.2 mm rainfall, and wide range of air temperature (maximum 35.0°C, minimum 15.6°C). Rice is the main crop grown over different land terrains and seasons. Aus (spring paddy), sesame and green gram in pre-kharif (early wet season), jute and *aman* rice in kharif (wet season) and wheat, different oilseeds and pulses, and potato in rabi (winter season) are important crops. South 24 Parganas (22°32'N to 22.53°N & 88°20'E to 88.33°E) has an area of 8165.05 km^2 with population of 81,53,176 (sixth highest in India), of which 74.39% stay in rural areas. Percentage of households below the poverty line is 37.21, much higher than the state and country average (26% and 29%). The district is one of the poor districts of West Bengal having large number of resource poor farmers. Natural resource is fragile and highly prone to degradation.

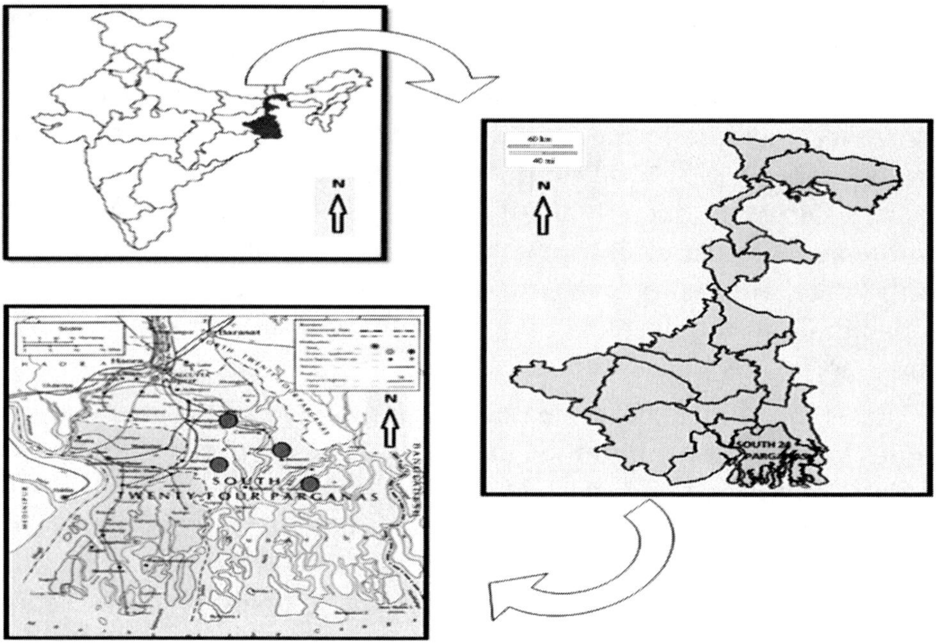

Figure 1: Study locations in South 24 Parganas district of West Bengal, India (shown by red circles).

Joynagar II, Baruipur, Basanti and Gosaba, the sampled Community Development Blocks, are distributed across the middle and eastern part of the district (Figure 1). Percentage of cropped and irrigated area of the district was 393.47 thousand ha and 98.31 thousand ha in 2006-07. Net cropped area of the district is 3,72,290 hectares, decreasing slowly due to the conversion of agricultural land for other purposes. Production of food grain, pulse and oilseed, vegetables and fruits has registered a growth in recent years (GoWB [2005]).

The district is characterized by complex, diverse and risk-prone ecosystem, which is perturbed by extreme climatic events, water logging in low lands for considerable part of the year and soil salinity in winter and summer months resulting in dearth of appropriate technology for the district. Population density has risen rapidly over the last decade, with 819 people per km^2 arable land (2011 Census) and at a growth rate of 18.05% between 2001 and 2011 (GoWB [2005]). Sustainable use of natural resources in the face of high population density and ecosystem fragility is critically important for the region. Overexploitation of natural resources is overwhelming in this district.

Farming has diversified greatly over recent years (Ghosh and Kuri [2005]) and technological intervention with appropriate technology has become critically challenging to the extension system. That is why classification of farming systems is a pragmatic starting point for facilitating appropriate technology in this region.

Sampling[a]

Following the guidelines of Indian Agricultural Statistical Research Institute, data collection was organized using a multi-stage random sample survey method (Sukhatme et al. [1984]). Four blocks (namely Joynagar II, Baruipur, Basanti and Gosaba) were randomly selected from 29 blocks of South 24 Parganas district. Then, three villages from each of the block and 12 farmers from each of the village were selected randomly. Care was taken that four farmers each were selected from marginal-small (2.50-7.50 bigha), medium (7.50-12.00 bigha) and large (>12.00 bigha) categories (1 bigha = 0.13 ha). A quick survey of these 144 households was conducted with a focus on socio-economic information and income of the farms from different farm enterprises. This data was used for identification of predominant farm types for the region. After the identification of predominant farm types, half of the members from each of the identified farm types were selected randomly and were interviewed in details for eliciting additional socio-economic and farm management related information (to be used for characterization). Thus, a total of 72 farmers were interviewed in details for characterization of farm types. This helped us to avoid detailed and time-consuming interviewing with all 144 respondents.

[a]Logistic support during sampling and data collection was received from All India Coordinated Research Project on Integrated Farming System and academic research programme of Ramakrishna Mission Vivekananda University.

Data

Instead of conducting a conventional way of identifying and characterizing farming system based on the source of the highest gross income received (from crops, vegetables, fruits and orchard, livestock, dairy, fishery, and poultry), the study employed multivariate statistics for identification of farm types and their characterization. Data were collected in two phases as mentioned in Sampling section.

Based on an extensive literature survey (Dixon et al. [2001]; Fliegel et al. [1968]; Mohammad[1978]; Singh et al. [1993]), initial case explorations in the field and expert counsel, a structured interview schedule was developed for the study. Apart from socio-economic parameters of the households, farm size, infrastructural facilities, information on cost of cultivation and yield of different crops, and price received by the farmers were included in the data collection instrument. The draft interview schedule was then pre-tested on 20 non-sampled respondents to incorporate necessary modifications. Description and measurement of some of the variables used in the study is given in Table 1.

Table 1: Selected variables used in farm type identification and characterization

Variable	**Description**
Education Index	Weighted average of formal education received by the household members (illiterate = 0, primary = 1, middle = 2, secondary = 3, higher secondary = 4, graduate = 5 and post-graduate = 6).
Crop Diversification	Measured by Herfindahl Index; index varies from 0 to 1, '1' - complete specialization and '0' - perfect diversification (Malik and Singh [2002])
System Cost of Cultivation	Cost of all inputs used (in Indian Rupees), for all enterprises of the farming system including the cost of family labor
System Gross Return	Total revenue earned (in Indian Rupees) from all the components of the farming system
System Net Return	System Gross Return – System Cost of Cultivation
Cost Benefit Ratio	Gross return from sale of output/Total cost of input used

Goswami *et al.*

Goswami *et al. Agricultural and Food Economics* 2014 2:5, doi:10.1186/s40100-014-0005-2

Data were collected during May-June 2012 through personal interview method from 144 sampled households for quick survey in the first phase followed by a detailed survey on 72 respondents in the second phase. The dataset was prepared and analyzed using SPSS 17 (SPSS Version 17, SPSS Inc., Chicago, Illinois) and XLSTAT 2012 software (Version 2012, Addinsoft, Inc., Brooklyn, NY, USA).

Statistical Analysis

Both univariate descriptive statistics and multivariate statistical techniques were employed for the analysis of data. Multivariate statistical techniques have been widely used for farm typology and characterization study (Kobrich et al. [2003]; Andersen et al. [2009]; Guto et al. [2010]). These statistical techniques allow us creating farm typologies, particularly when an in-depth database is available. The combination of principal component analysis for necessary reduction of the number of variables followed by cluster analysis to identify typical farm households was used. Similar examples may be drawn from Gebauer ([1987]) in Germany that employs cluster analysis; Jolly ([1988]) in Senegal that employs canonical-discriminant analysis for identification of recommendation domains; Orr and Jere ([1999]) in Malawi employing cluster analysis followed by profiling through discriminant analysis; Hardiman et al. ([1990]) in China using cluster analysis; Köbrich et al. ([2003]) in Chile and Pakistan and Usai et al. ([2006]) in Sardinia, both using principal component analysis and successive cluster analysis.

For the present study, farm household typologies were identified by using two sequential multivariate statistical techniques: principal component analysis (PCA) and cluster analysis (CA) (Ding and He[2004]). PCA condenses the information from the interdependent variables to a smaller set of factors (Jolliffe [2002]; Abdi [2007]). We used both socio-economic variables and income from different crop enterprises in the PCA. Since the impact of off-farm income on technology adoption is well-reported, the present study also used off-farm income as a factor for classification (Savadogo et al. [1998]; Nehring et al. [2005]; Fernandez-Cornejo et al. [2005]).

Factors were identified using orthogonal rotation (varimax method as in Kaiser [1970]; Gorsuch[1983]) so that a smaller number of highly-correlated variables might be put under each factor and interpretation becomes easier (Field [2005]). In accordance with Kaiser's criterion, all factors exceeding an eigenvalue of one were retained (Kaiser [1970]). Kaiser's criterion is accurate when the number of variables is less than 30 (Field [2005]), which was the case for our data set.

The sampled farms were clustered based on the five principal components identified by PCA. Cluster analysis is a collective term covering a wide variety of techniques for defining natural groups or clusters in data sets (Anderberg [1973]). These groups are relatively homogeneous within themselves and heterogeneous between each other based on a defined set of variables. In a word, it is the art of finding groups in data (Kaufman and Rousseeuw [1990]). Entities within a certain group or cluster are expected to be very similar to each other, and entities belonging to different classes are expected to be dissimilar (Bidogeza et al. [2009]).

To determine the number of clusters two steps were followed for clustering – the hierarchical method and K-means clustering method. For hierarchical clustering, Euclidian distance and Ward's computation method was considered. The number of clusters retained from Ward's method (four in our study) was used as starting values in the K-means method. Accordingly, the number of clusters that seemed most realistic and meaningful was chosen for the final solution.

In addition to CA, a one-way analysis of variance test was performed. The test allowed us to identify the differences in variance between clusters in terms of PC scores (Field [2005]). Thus, the variables that explained the largest differences between clusters could be identified.

RESULTS AND DISCUSSION

Description of the Farming Systems Four major types of farms, based on the source from which maximum gross income was earned by the farmers, were identified. These were – rice-based farming system (34 households), vegetable-based farming system (70 households), fishery-based farming system (10 households) and farming system based on off-farm income (20 households). The farm size wise distribution of these farm types in four study Blocks has been reported in Figure 2.

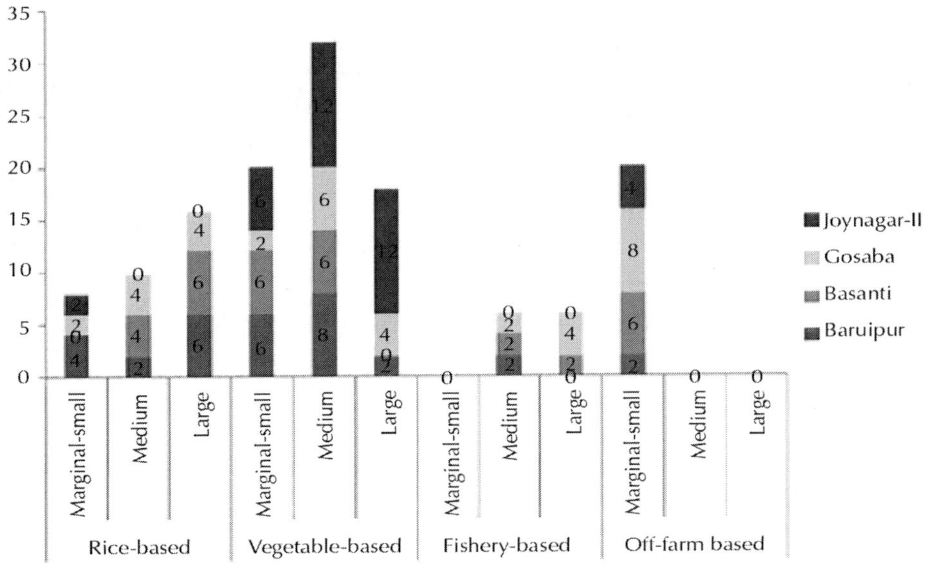

Figure 2: Farmer category wise sub-farming systems in four study blocks of South 24 Parganas district.

A general observation revealed that large proportion of farmers followed rice-based farming systems, perhaps to grow rice for self-consumption and not for sale. On the other hand, medium farmers followed vegetable-based farming systems and marginal-small farmers relied on off-farm income sources. This might be due to the pressing need for income from their small holdings. Medium farmers could allocate enough land for vegetables after growing rice for subsistence purposes. However, vegetable-based farming systems were found across farm sizes, perhaps for their high economic return. Chili, tomato, brinjal, ladies finger, and cole crops were common in this part of the state (Basu et al. [2009]; Halder and Das [2012]). Fishery-based farming systems were found among medium and large farmers only, who could afford to spare land for water bodies, in case of dug pond, and invest in commercial fishery. Spatially, most of the vegetable-based systems were found in Joynagar-II Block followed by Baruipur Block. In Basanti and Gosaba Block, all types of farm were found. A considerable proportion of these farms earned large off-farm income.

Had we used the conventional method for identifying farm types, i.e. classification based on farm size or income sources of the farm, the diversity of farming systems would have been grossly underestimated.

For example, it is difficult to understand the importance of fishery (only 10 households) and animal husbandry (only four households, not featured in Figure 2) in different farming systems if we go by the numbers in Table 2 only.

Table 2: Predominant sub-farm types under each farm types in the study blocks; sub-farm types are arranged according to the decreasing contribution of an enterprise towards gross farm income

Farming systems	Sub-farming systems			
	Joynagar-II	**Baruipur**	**Basanti**	**Gosaba**
Rice-based	Rice + Off-farm + Vegetables + Livestock + Pulses + Fishery + Fruit + Poultry	Rice + Vegetables + Off-farm + Livestock + Fishery + Poultry	Rice + Vegetables + Off-farm + Livestock + Fishery + Fruit + Goatery + Poultry	Rice + Vegetables + Off-farm + Livestock + Fishery + Goatery
		Rice + Vegetables + Fishery + Livestock + Plantation + Fruit + Poultry	Rice + Vegetables + Fishery + Livestock + Goatery + Plantation + Fruit	Rice + Vegetables + Off-farm + Fishery + Livestock + Fruit
		Rice + Poultry + Vegetables + Fishery + Plantation + Livestock	Rice + Vegetables + Off-farm + Fishery + Livestock + Fruit + Goatery	Rice + Fishery + Vegetables + Livestock + Goatery + Fruit + Poultry
		Rice + Vegetables + Off-farm + Livestock + Fishery + Tree + Poultry	Rice + Vegetables + Off-farm + Fishery + Livestock + Fruit	Rice + Off-farm + Fishery + Livestock + Goatery + Poultry
		Rice + Fruit + Off-farm + Vegetables + Livestock + Fishery + Plantation	Rice + Vegetables + Off-farm + Fishery + Livestock	Rice + Livestock + Vegetables + Fishery + Farm Machinery + Goatery + Poultry
		Rice + Vegetables + Off-farm + Fruit + Livestock + Fishery		

Vegetables-based	Vegetables + Rice + Livestock + Pulses + Fishery + Goatery + Tree + Poultry	Vegetables + Rice + Off-farm + Fishery + Livestock + Oilseed	Vegetables + Rice + Fishery + Off-farm + Livestock + Fruit + Goatery + Poultry	Vegetables + Rice + Livestock + Fishery + Goatery + Fruit crop
	Vegetables + Rice + Pulses + Fruit + Fishery + Livestock + Poultry	Vegetables + Rice + Fruit + Livestock + Fishery + Goatery	Vegetables + Rice + Off-farm + Fishery + Fruit + Livestock + Plantation + Goatery	Vegetables + Rice + Livestock + Fishery + Fruit + Goatery + Poultry
	Vegetables + Rice + Fruit crop + Pulses + Livestock + Oilseed + Fishery + Plantation + Poultry	Vegetables + Rice + Fruit crop + Fishery + Livestock	Vegetables + Fishery + Rice + Livestock + Poultry + Fruit	Vegetables + Rice + Fishery + Fruit crop + Fishery + Livestock
	Vegetables + Rice + Fruit + Fishery + Goatery + Pulses + livestock	Vegetables + Off-farm + Rice + Livestock + Goatery + Poultry	Vegetables + Rice + Fishery + Plantation + Goatery + Fruit + Poultry	Vegetables + Rice + Fishery + Livestock + Tree + Poultry
	Vegetables + Rice + Fruit + Pulses + Fishery + Livestock + Oilseed + Plantation	Vegetables + Poultry + Fruit + Rice + Livestock + Fishery + Pulses	Vegetables + Rice + Off-farm + Fishery + Livestock + Poultry	Vegetables + Rice + Off-farm + Livestock + Fishery + Goatery
	Vegetables + Rice + Pulses + Fruit + Fishery + Livestock + Tree + Poultry	Vegetables + Rice + Fishery + Goatery + livestock	Vegetables + Off-farm + Rice + Fishery + Livestock + Poultry	Vegetables + Rice + Fishery + livestock + Poultry + Fruit crop
	Vegetables + Rice + Fruit + Plantation + Pulses + Fishery + Livestock	Vegetables + Rice + Fruit + Fishery + Livestock + Goatery		
	Vegetables + Rice + Oilseed + Goatery + Fruit + Fishery + Livestock + Poultry	Vegetables + Rice + Off-farm + Goatery + Fishery + Tree + Fruit + Poultry		
	Vegetables + Rice + Fruit + Fishery + Pulses + Plantation + Livestock			
	Vegetables + Rice + Fishery + Goatery + Pulses + Plantation			
	Vegetables + Rice + Livestock + Pulses + Fishery + Tree + Fruit crop			
	Vegetables + Rice + Pulses + Fruit + Livestock + Plantation + Fishery + Oilseed			
	Vegetables + Rice + Livestock + Fishery + Tree + Goatery + Poultry			
	Vegetables + Fishery + Rice + Fruit + Pulses + Plantation + Livestock + Cotton			
	Vegetables + Rice + Off-farm + Goatery + Fishery + Livestock + Pulses + Plantation			

Fishery-based		Fishery + Vegetables + Cereal + Fruit + Livestock + Oilseed	Fishery + Cereal + Livestock + Goatery + Poultry	Fishery + Vegetables + Rice + Livestock + Goatery + Fruit + Poultry
			Fishery + Cereal + Vegetables + Livestock + Poultry	Fishery + Rice + fruit crop + Livestock + Poultry
				Fishery + Rice + Vegetables + Off-farm + Fruit crop + Livestock + Farm Machinery + Tree + Poultry
Off-farm based	Off-farm + Cereal + Vegetables Livestock + Fruit + Poultry + Fishery + Tree	Off-farm + Cereal + Livestock + Fishery + Poultry	Off-farm + Rice + Livestock + Plantation + Tree + Fishery + Fruit	Off-farm + Rice + Fishery + Livestock + Poultry
	Off-farm + Vegetables + Cereal + Livestock + Fishery + Tree + Poultry		Off-farm + Rice + Livestock + Fishery + Fruit + Poultry	Off-farm + Rice + Livestock + Poultry + Goatery + Fruit
			Off-farm + Rice + Livestock + Vegetables + Fishery + Poultry + Fruit	Off-farm + Vegetables + Rice + Livestock + Poultry + Fishery + Tree
				Off-farm + Rice + Livestock + Vegetables + Fishery + Poultry

Goswami *et al.*

Goswami *et al.* *Agricultural and Food Economics* 2014 2:5, doi:10.1186/s40100-014-0005-2

In all the four study blocks, a large number of sub-farming systems were found. These sub-farming systems were determined by unique combination of farm enterprises (Table 2). The label of enterprises in a given sub-system was given according to their contributions to the gross farm income. For example, 'Rice + Jute + Vegetables' sub-system denoted that Rice contributed highest to the farm income, followed by Jute and Vegetables. Number of sub-farming systems in Joynagar-II, Baruipur, Basanti and Gosaba Blocks were 18, 16, 16 and 18 respectively. This huge diversity in sub-farming systems (68 in no.) was inherent to the farms of this region and they ideally required specific extension interventions. However, in practice, it was difficult to address the needs of all these 68 sub-farming systems.

Since these sub-systems are huge in number and farm size or enterprise-based classification is oversimplification of farm heterogeneity, we followed an alternative methodology for identification of farm types and their characterization by employing a combination of PCA and CA with the derived principal components. The results of these multivariate analyses are described below.

The Principal Component Analysis

Table 3 shows rotated factor (Varimax) matrix of independent variables with factor loadings for each variable. The communality column shows the total amount of variance of each variable retained in the factors. For the interpretation of the PCs, variables with high factor loadings and high communality were considered from the rotated factor matrix (Harris [2001]). In total, 17 variables were included in PCA, of which 5 principal components with eigenvalues greater than 1 were retained for further analysis. These five PCs explained 73.46% of total variability in the dataset. A closer look at each column of Table 3 helps us to define each component according to the strongly associated variables. The first component explains 27.78% variance and is correlated with the land holding, owned land, land fragment and income from rice crop. Thus the component represents land entitlement and income from rice crop. Principal component 2, 3, 4 and 5 explain 15.87%, 11.02%, 9.59% and 8.20% variance respectively. That means the first two factors have more importance in explaining the variance in dataset than the other three factors combined. PC 2 is correlated with crop diversification index and income from pulse, oilseed, fruits and plantation, representing crop and income diversification. PC 3 is correlated with income from poultry, fishery and family size, embodying the income from poultry and fishery enterprises driven by family labor. PC 4 represents off-farm income and PC 5, correlated with income from goatery and cattle, represents livestock enterprises. Thus the five principal components could be names as 'land ownership and income from staple food' (PC 1), 'farm diversification' (PC 2), 'income from poultry and fishery' (PC 3), 'off-farm income' (PC 4), and 'income from livestock' (PC 5).

Table 3: Four principal components derived by principal components analysis with loadings for individual crop enterprises and percent cumulative variance explained

	PC 1	PC 2	PC 3	PC 4	PC 5	Communality
Family Size	0.33	0.25	0.55	0.34	0.16	0.62
Land Holding	0.89	0.15	0.19	0.14	0.10	0.88
Own Land	0.69	0.16	0.14	0.19	0.17	0.60
Land Fragment	0.80	0.04	0.02	0.05	0.04	0.65
Crop diversification Index	0.22	−0.69	−0.07	0.26	0.05	0.61
Income from Rice	0.84	−0.12	0.09	−0.15	0.18	0.79
Income from Pulses	0.10	0.78	−0.16	0.34	−0.09	0.77
Income from Oilseed	−0.03	0.59	0.19	−0.06	0.33	0.54
Income from Vegetables	0.46	0.25	−0.21	0.47	0.20	0.56
Income from Fruits	0.37	0.75	0.14	0.04	−0.17	0.74
Income from Tree	−0.27	−0.10	−0.09	0.10	−0.47	0.53
Income from Plantation	0.28	0.63	−0.08	0.30	0.06	0.58
Income from Cattle	0.25	0.12	−0.07	0.10	−0.55	0.63
Income from Poultry	0.09	−0.04	0.92	0.00	0.01	0.86
Income from Fishery	0.41	−0.15	0.65	−0.02	−0.14	0.70

Income from Goatery	0.01	−0.15	−0.02	0.19	0.85	0.77
Off-farm income	0.07	−0.13	−0.19	−0.83	−0.08	0.79
Eigen values	4.72	2.36	1.36	1.29	1.05	
Cumulative explained variance	27.78	43.66	54.68	63.26	73.46	

Goswami *et al.*

Goswami *et al. Agricultural and Food Economics* 2014 2:5, doi:10.1186/s40100-014-0005-2

The Cluster Analysis

The first five factors from the principal component analysis were used for hierarchical clustering using Euclidean Distance as distance measure and Ward's technique as agglomerative clustering. The agglomeration schedule, resulting from this analysis, illustrated the sequence of analysis and produced coefficients. The objective of generating such schedule is to arrive at an appropriate number of clusters that best fit the data set. A check of the agglomeration schedule and Scree Diagram suggested that the number of clusters should be four. It was then checked that the number of retained clusters is realistic with respect to the field observation in order to be accepted as a meaningful classification. Thus, four clusters were found to be appropriate, as these seemed to be most representative of farm households within the study blocks. The k-means clustering method was employed with 4 numbers of clusters. The analysis produced final cluster centres (Table 4), and distribution of farms within a cluster among study blocks (Table 5) and size of farms (Table 6).

Table 4: Final cluster centres for all four clusters identified through K-means clustering method

Principal components	**Clusters**			
	1	**2**	**3**	**4**
PC 1 (Land ownership and income from cereals)	0.84	0.23	−0.31	0.53
PC 2 (Farm diversification)	−0.09	0.87	−0.17	0.53
PC 3 (Income from poultry and fishery)	3.58	0.10	−0.19	0.50
PC 4 (Off-farm income)	−0.03	−0.20	−0.72	0.35
PC 5 (Income from livestock)	−0.00	−0.50	−0.07	2.14

Goswami *et al.*

Goswami *et al. Agricultural and Food Economics* 2014 2:5, doi:10.1186/s40100-014-0005-2

Table 5: Distribution of farm types, identified by cluster analysis, in four study blocks of South 24 Parganas districts

	Farm types				Total
	Cluster 1	**Cluster 2**	**Cluster 3**	**Cluster 4**	
Joynagar-II	0 (0.00, 0.00)*	18 (36.00, 50.00)	10 (14.29, 27.78)	8 (57.14, 22.22)	36
Baruipur	8 (100.0, 22.22)	8 (16.00, 22.22)	18 (25.71, 50.00)	2 (14.29, 5.56)	36
Basanti	0 (0.00, 0.00)	8 (16.00, 23.53)	22 (31.43, 64.71)	4 (28.57, 11.76)	34
Gosaba	0 (0.00, 0.00)	16 (32.00, 44.44)	20 (28.57, 55.56)	0 (0.00, 0.00)	36
Total	8	50	70	14	142
Likelihood ratio	46.13	Sig. = .000			

*Figures in parentheses are percentages to column and row marginal respectively.

Goswami *et al.*

Goswami *et al. Agricultural and Food Economics* 2014 2:5, doi:10.1186/s40100-014-0005-2

Table 6: Distribution of farm types, identified by cluster analysis, among three farmer categories in South 24 Parganas districts

	Farm types				Total
	Cluster 1	**Cluster 2**	**Cluster 3**	**Cluster 4**	
Marginal-small	0 (0.00, 0.00)*	4 (8.00, 8.33)	44 (82.86, 91.67)	0 (0.00, 0.00)	48
Medium	2 (25.00, 4.35)	26 (52.00, 56.52)	14 (20.00, 30.43)	4 (28.57, 8.70)	46
Large	6 (75.00, 12.5)	20 (40.00, 41.67)	12 (17.14, 25.00)	10 (71.43, 20.83)	48
Total	8	50	70	14	142
Likelihood ratio	67.089	Sig. = .000			

*Figures in parentheses are percentages to column marginal and row marginal respectively.

Goswami *et al.*

Goswami *et al. Agricultural and Food Economics* 2014 2:5, doi:10.1186/s40100-014-0005-2

Final cluster centers interpret what is typical for a particular cluster. Cluster 1 is characterized by relatively higher income from poultry and fishery enterprises, has large holdings, and earns high from rice crop. Cluster 2 is characterized by higher diversification of farm income and Cluster 3 is characterized by smaller farms with higher off-farm income. Cluster 4 represents farms with higher income from livestock and other diverse sources such as rice, vegetables, poultry and fishery (Table 4). In all clusters, the contribution of several important crops and off-farm income could be observed.

Figures 3 & 4 (A, B) represents the observations and variables simultaneously in the two dimensional space. The individual farms have been labeled with study blocks (Figure 3A & B) and size category of farms (Figure 4A & B). The A series (3A & 4A) depicts the distribution of observations in a space defined by two principal components, after varimax rotation. The B series (3B & 4B) depicts the same along with the initial variables plotted in the form of vectors. Farms clustered (circled) together represent a farm type. A visual check suggests one group at the right and lower quadrant of the figure (Figure 3A), another clustered at the left lower quadrant, which is closer to the axis. The third one is also located at the left lower quadrant, but away from vertical axis. The scattered farms of upper right and left quadrants form the fourth group. A check of these Figures (3 & 4) in combination with Tables 5 and 6 helps us to understand this clustering better.

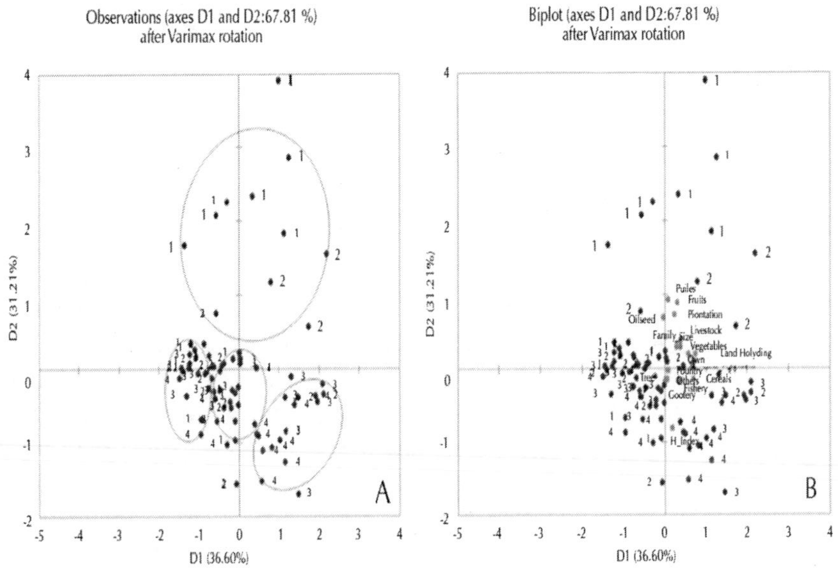

Figure 3: Distinct cluster of farms delineated by PCA axis and labeled by study blocks.A. Principal component analysis observations plot diagram of farming systems based on PCA of variables after varimax rotation. The numbers 1-4 on the plane indicate the community development blocks namely Joynagar II - 1, Baruipur - 2, Basanti - 3 and Gosaba - 4. The distance between the positions of the farms approximates the dissimilarity of their variables measured by Euclidean distance. Four circles/ovals represent four farm types. B. Ordination bi-plot diagram of farming system characterization based on PCA of variables. The numbers 1-4 on the plane indicate the community development blocks namely Joynagar II - 1, Baruipur - 2, Basanti - 3 and Gosaba - 4. Red dots represent variables used in PCA. The distance between

the positions of the farms approximates the dissimilarity of their variables measured by Euclidean distance. Samples close to the origin have average values of a particular variable in a study sample.

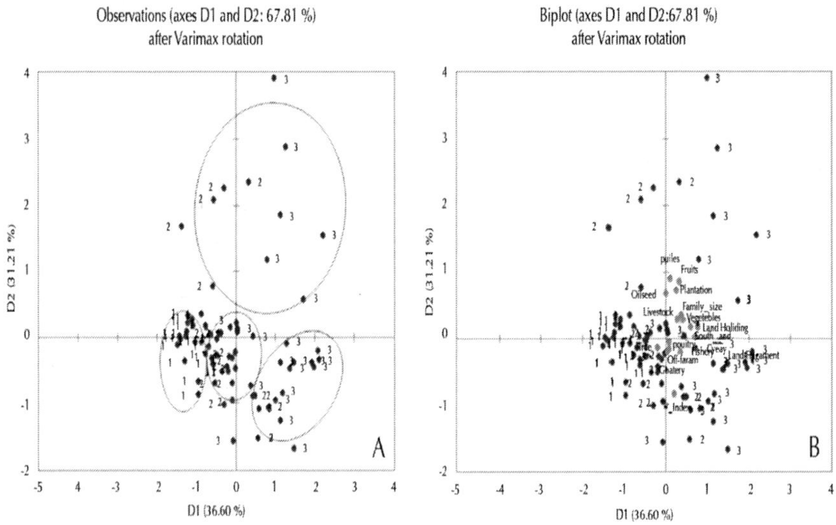

Figure 4: Distinct cluster of farms delineated by PCA axis and labeled by size category of farms.A. Principal component analysis observations plot diagram of farming systems based on PCA of variables after varimax rotation. The numbers 1-3 on the plane indicate the different farm sizes – Marginal-Small - 1, Medium – 2, and Large - 3. The distance between the positions of the farms approximates the dissimilarity of their variables measured by Euclidean distance. Four circles/ovals represent four farm types. B. Ordination bi-plot diagram of farming system characterization based on PCA of variables. The numbers 1-3 on the plane indicate the different farm sizes – Marginal-Small - 1, Medium – 2, and Large - 3. Red dots represent variables used in PCA. The distance between the positions of the farms approximates the dissimilarity of their variables measured by Euclidean distance. Samples close to the origin have average values of a particular variable in a study sample.

Noticeably, type of farms is not specific to a study block. The labels '1' , '2' , '3' and '4' are scattered throughout the plot. However, in some places similar labels are clustered signifying similar farms within a Block. The findings from Figure 4A are even more interesting. Small and marginal farms labeled as '1' are clustered together signifying their similarity in terms of a host of factors. Large farms labeled as '2' are also found closely clustered. Instead, medium farms are scattered throughout the plot representing higher instrinsic variability of medium farms. The nearness of blue dots (individual farms) to the red dots (initial variables) shows the characteristics of those farms in terms of the variables (represented by red dots). A close look of both 3A and 4A provides better understanding of farm classification in the study areas.

It is observed that all the Cluster 1 farms (100%) are situated in Baruipur Bock (Table 5). Cluster 2 farms are predominant in Joynagar II (36.00%) and Gosaba (32.00%), Cluster 3 farms are distributed across the four study blocks and Cluster 4 farms are predominant (57.14%) in Joynagar II Block. The data suggest that Joynagar II Block is predominated by diversified farms, Baruipur and Basanti Blocks by smaller farms with higher off-farm income and Gosaba Block with diversified and smaller farms with off-farm income. The distribution of clusters across the study blocks differed significantly (Likelihood ratio sig. value = .00) indicating heterogeneous nature of farm types in these blocks.

From Table 6, it may be observed that marginal-small farms are mostly Cluster 3 farms (91.67%), i.e. smaller farms with higher off-farm income. Medium farms are Cluster 2 (56.52%) and Cluster 3 (30.43%) farms i.e. diversified farms and small farms with higher off-farm income. Noticeably, large farms are distributed across all clusters, perhaps due to their higher flexibility in allocation of land resources among different crops/enterprises. Alternatively, Cluster 1 farms are mostly (75.00%) large farms, Cluster II farms are largely Medium (56.52%) and Large (41.67%) farms, Cluster 3 farms are mostly Marginal-small farms (82.86%) and Cluster 4 farms are Large farms (71.43%). The distribution of clusters across the farm size differed significantly (Likelihood ratio sig. value = .00) indicating heterogeneous nature of farm types across farm sizes.

In some parts of West Bengal including South 24 Parganas district, departure from traditional cropping pattern has been a trend for the last one and half decades (Goswami [2007]; Bhattacharyya[2008]) for enhancing income from fragmented land resources, which are challenged by abiotic stresses (e.g. salinity). The number of smallholders has in fact increased in West Bengal (De [2003]) and there is evidence of increased crop diversity on fragmented lands in the state (De [2002]; Bagchi et al. [2012]). In many places, farmers have changed land shapes to grow vegetables in low lands (Basu et al.

[2009]). Introduction of farm-ponds and land-shaping has also resulted in high cropping and income diversity among small farmers (Bahu and Mishra [2001]). Off-farm income has also become a burgeoning reality of rural India especially for the smallholders and this has often become the largest source of rural farm income. This also contributes to the investment in farming activity and asset creation (Mehta [2009]).

To examine the authenticity of the cluster analysis, a one-way analysis of variance for each principal component (equality of group means) was performed with the four clusters (Table 7). The more distinct a PC value is among groups, the lower is the p-value. Judging from the p-values, all the factors namely 'land ownership and income from rice crop' , 'income from poultry and fishery' , 'off-farm income' and 'income from livestock' found to be significant in differentiating the clusters (p = 0.00 for all factors). However, factor 2 characterized by 'farm diversification' was not significantly differing across identified clusters, although the p value was not far from significant (p = 0.09).

Table 7: One-way analysis of variance of factor scores across four identified clusters

Factors	Cluster		Error		F	**Sig.**
	Mean square	**df**	**Mean square**	**df**		
PC 1 (land ownership and income from rice crop)	4.68	3	.92	138	5.10	0.00
PC 2 (farm diversification)	2.14	3	.98	138	2.20	0.09
PC 3 (income from poultry and fishery)	26.06	3	.46	138	57.25	0.00
PC 4 (off-farm income)	36.30	3	.23	138	156.12	0.00
PC 5 (income from livestock)	25.58	3	.47	138	54.95	0.00

Goswami *et al.*

Goswami *et al. Agricultural and Food Economics* 2014 2:5, doi:10.1186/s40100-014-0005-2

Characterization of Identified Farm Types

Table 8 helps us to understand the characteristics of the clusters in terms of background variables and economic performance indicators of the farming systems.

Table 8: Characteristics of identified clusters of farm households and p-value of one way analysis of variance (equality of group mean) for variables used in characterization of farm types

	Cluster 1	Cluster 2	Cluster 3	Cluster 4	P-value
	(N = 4)	(N = 25)	(N = 35)	(N = 7)	
	(1)	(2)	(3)	(4)	
Age (yrs)	50.50[ab]**	51.68[a]	46.03[b]	53.86[a]	0.00
Family size	9.50	6.72	5.60	7.29	0.00
Farm size (bigha)*	10.38[a]	7.71[bc]	5.28[c]	9.65[ab]	0.00
Own land (bigha)	6.88[ab]	6.55[b]	4.41[b]	9.65[a]	0.00
Leased in land (bigha)	3.50[a]	1.16[b]	0.87[bc]	0.00[c]	0.00
Leased out land (bigha)	0.28[a]	0.12[b]	0.07[b]	0.36[a]	0.02
Land fragment (no.)	4.25[ab]	3.96[b]	3.22[c]	4.43[a]	0.01
Distance from home	2.0[a]	1.08[b]	1.08[b]	1.28[b]	0.00
Education Index	2.27	1,92	1.92	1.97	0.20
Area of Rice (bigha)	8.00[a]	6.16[ab]	4.76[b]	7.79[a]	0.00
Area of Vegetables (bigha)	0.75[b]	1.40[a]	0.70[b]	1.84[a]	0.00
Area of oilseeds (bigha)	0.23[b]	0.86[ab]	0.09[b]	1.43[a]	0.00
Crop Diversification Index	0.29[ab]	0.33[a]	0.28[b]	0.28[b]	0.01
System Gross return (Rs.)	139640[ab]	110980[b]	66508[c]	149550[a]	0.00
System Cost of Cultivation (Rs.)	60391[ab]	47228[b]	30330[c]	64218[a]	0.00
System Net Return (Rs.)	79250[a]	63755[a]	34576[b]	79449[a]	0.00
Cost-Benefit Ratio	2.42	2.34	2.39	2.28	0.88
Income from Rice (Rs.)	37380[a]	23165[b]	20734[b]	34563[a]	0.00

Income from pulses (Rs.)	576^{c}	3538^{ab}	310^{bc}	4071^{a}	0.00
Income from Oilseed (Rs.)	1050^{b}	296^{b}	40^{b}	2657^{a}	0.00
Income from Vegetables (Rs.)	23610^{b}	42148^{a}	16817^{b}	46093^{a}	0.00
Income from Fruits (Rs.)	9700^{a}	6584ab	2986^{2}	5185^{b}	0.01
Income from plantation (Rs.)	1750^{b}	1856b	332^{c}	3614^{a}	0.00
Income from Livestock (Rs.)	5600^{ab}	6059^{a}	4738^{b}	6558^{a}	0.01
Income from poultry (Rs.)	50125^{a}	503^{b}	701^{b}	212^{b}	0.00
Income from Fishery (Rs.)	8250^{b}	18816^{a}	4995	15800^{a}	0.00
Income from goatery (Rs.)	1600^{b}	650^{c}	1145^{b}	6485^{a}	0.00
Off-farm Income (Rs.)	0^{c}	800^{c}	15171^{a}	2857^{b}	0.00

*1 bigha = 0.133 ha.

**Difference in mean among clusters denoted by different letters as found from Duncan Multiple Range Test. Highest mean value is denoted by a, followed by b and c when significant mean difference was found between clusters.

Goswami *et al.*

Goswami *et al. Agricultural and Food Economics* 2014 2:5, doi:10.1186/s40100-014-0005-2

Cluster I accounts for 5.63% of farm households. The cluster is comprised of households having large land holding, large family size (representing joint extended families), relatively higher family education, and relatively higher crop diversification (Col. 1, Table 8). The households both lease out and lease in land, land are highly fragmented and situated at a high distance from the dwellings. In terms of economic performance indicators, this cluster is characterized by relatively high system gross return, higher cost of cultivation and system net return and relatively higher cost-benefit ratio. These households, on an average, secured high income from rice crop, poultry and vegetables followed by fruits and fishery. They depended wholly on agriculture with no off-farm income. These farms may be supported for technically sound intensification of agriculture with assured input and advisory services. Since these groups pursue a capital intensive diversified farming, access to credit is important for them. The farmers may be assisted in commercial production techniques with provision for building requisite capacity in farm and non-farm based entrepreneurship (Folmer et al. [2010]). Export-oriented crops may be tried with these farmers with suitable incentive (Singh [2002]). Since these farms are capital and management intensive, they are prone to become environmentally unsustainable in the long run. An environmental monitoring mechanism is needed to be in place in addition to an incentivized system in favour of sustainable farming (Dhiman and Thambi [2009]).

Cluster II comprised of 35.21% of the farm households. The cluster members had relatively smaller land, smaller family size, relatively poor family education, and high crop diversification (Col 2, Table 8). These farms lease-in land and their fragmented lands are relatively nearer to their settlement. This cluster is characterized by low system gross return, low cost of cultivation, moderate system net return, but relatively better cost-benefit ratio. These households secured large proportion of income from vegetables, fishery, livestock and rice crop and may be grouped as diversified farms. These farms are management intensive, presumably supported by human labor, and pursue diversification in enterprises. Integrated farming with livestock and fishery will help in higher food and livelihood security without undermining farm resources (Prein [2002]). This will require specialized extension support with a farming system approach, and institutional convergence for sound planning (Nabi [2008]). Cash crops may be promoted for enhancing farm income with assured provision for credit and insurance (Fafchamps [1992]). On account of the large proportion of such farms, collectivization might be a viable strategy for their market integration (Hazell [2005]).

Cluster III comprises of 49.30% of the sampled farms. This cluster is characterized by small land holding, relatively poor family education, smaller family size and relatively low crop diversification (Col. 3, Table 8). These farms have started shifting to off-farm income to offset low income from farming. In terms of economic performance, this cluster is characterized by low system gross return, low cost of cultivation or low investment in farming, poor net return from farming, but higher benefit-cost ratio (since they invest less). The farm households earn most of their income from rice crop, vegetables and from off-farm sources. These are typical subsistence farms that have started to shift from traditional farming to non-farming vocations, often in the form of seasonal migration. Maintaining food security is a key policy goal for these farms, which may be achieved through sustainable and risk-averting cultivation of food grains (e.g. climate resilient agriculture) coupled with assured coverage under public distribution system for food grains (Thomas [2006]). This might be supplemented with a comprehensive capacity building programme on non-farm enterprises, since these people are poor in human resources and exploited in the labor markets in nearer towns (Deb et al. [2002]). Collectivization of small enterprises may follow this capacity building programme for their market integration.

Cluster IV comprises of 9.86% of the farm households. The cluster members had relatively larger land holding (owner cultivator), relatively larger family size, moderate family education, and moderate crop diversification (Col. 4, Table 8). They do not lease-in land, i.e. these farms represent owner cultivator who lease-out land occasionally. Land is highly fragmented. These households secure high income from rice crop, vegetables, and fishery and are characterized by high system gross return, high cost of cultivation, high system net return and relatively low benefit-cost ratio. The cluster members had little off-farm income. These farms are capital intensive, developed to meet growing food demand in the nearest town markets. Unlike Cluster I, these farms earn higher income from vegetables and fishery. These may be supported for intensive farming with improved technology. Since they specialize is in food grain and vegetables, assured marketing support and crop insurance are critical for these farms. Due to their available family labors, sustainable agriculture and agro-based enterprises also seem to be promising propositions (Bhattacharjee and Saravanan[2012]).

This farm typology classification offers clear advantages over classifications based on farm size or agro-ecological characteristics. The farm types delineated are manageable in number and represent both socio-economic, resource ownership and management orientation of the farms. Farm size-based classifications undermine the huge diversity among size classes and agro-ecological classification and ignore socio-economic realities of the farms. At the same time, the farm classes are based on sound statistical procedures instead of size-based classifications and, hence, more acceptable to policymakers.

The four farm types differed among themselves in terms of all studied parameters including a string of background variables, income sources and economic indicators, showing the efficiency of the classification methodology employed. The farm types, however, did not differ in terms of education index and cost-benefit ratio. Increasing dependence on costly external inputs has perhaps rendered an increase in cost of cultivation, relative to system net return, irrespective of farm types. Education was not an important means to classify farms, as access and control over natural resource was more important in differentiating farm types.

CONCLUSIONS AND IMPLICATIONS

The implications of the study may be illustrated in practical and academic terms. While the practical focus is on institutional arrangement and policy interventions, the academic focus has been on methodological insights gained and its possible upscaling.

The Methodological Implication and Scope for Modification and Upscaling

The study departs from conventional methodology of economic characterization of predominant farming systems (classification based on land holding or discussion with stakeholders with little statistical rigor) in terms of at least two dimensions. First, it uses multivariate statistics for identification of farm types, and second, it uses a string of economic and non-economic variables for such classification. This methodology has reduced the time of data collection significantly, as quick survey was employed for farm type identification and only half of the households were interviewed in detail for their economic characterization. In-depth and comprehensive classification of farms was possible resulting in a manageable number of farm types (four) that could be taken up for appropriate policy and technological intervention.

The methodology employed in this study can be modified under different situations of farm typology study. The variables used for PCA may be different, based on the nature of agro-ecosystem, nature of agriculture practiced, objective of classification (nutrient management, irrigation intervention, credit support) etc. Variables may also vary when the focus of farm characterization is different from economic characterization only (e.g. these may be energy efficiency, ecological sustainability). The implication would also reflect in the development of data collection instruments to be used in quick and detailed survey. The process of variable selection may involve stakeholder participation, thus promoting popular participation in the study. The multivariate approach (algorithm) followed for the farm type delineation may also be used for the development of software or web-based decision support system. Separate modules in the software may be incorporated with flexible set of indicators under each of them. This may be used by research and extension agencies after rigorous multi-locational field testing.

Proposed Institutional Environment for Technology Transfer

The results from the analysis suggested four major farm types in the study areas – (i) capital-intensive large diversified farms with high economic efficiency, (ii) management-intensive, medium, diversified farms with moderate economic efficiency, (iii) small, subsistence farms with high off-farm income and high economic efficiency, and (iv) large diversified owner-cultivated farms with relatively lower economic efficiency. Smaller farms have inclined towards non-farm income more than the medium and large farms. On the other hand, medium and large farms have diversified more than smaller farms with a departure from grain-based systems to vegetable-based farming systems. This asks for a holistic farm planning and extension intervention than confining efforts to technology transfer alone. India, along with a large number of countries having agrarian society, has entered the open economy regime and need to establish economic efficiency for their smallholder system (Birner et al. [2009]). The extension system must precisely target agricultural inputs, advisory services, credit access and critical information for identified farm. The selection of beneficiaries for many public extension programs may also be guided by such farm typology.

To respond to the need of such comprehensive farm types, institutional innovation (development and/or convergence) often becomes necessary. In India, a nationwide revitalized extension system has just started functioning. It explicitly encourages such convergence in planning and execution of technology transfer programs. Linkage with existing donor supported and Ministry funded efforts might also lead to improved institutional environment. India is now experiencing a handful of flagship rural development programs that builds on constitutional provisions for basic services to its citizens and development of rural institutions (mostly in the form of rural collectives), and these may be efficiently tapped for effective transfer of agricultural technologies.

Extrapolation of the Study Outcome

The study does not necessarily claim to extrapolate the identified farm types to larger context. But, since the study locations are representative of coastal saline zone of India, this may be generalized for at least 10 Mha of cultivable lands with nearly 900 million people operating under this system. However, it is the methodology that is subject to generalization for even larger regions. India has almost 200 districts under complex agro-ecological systems covering a population of nearly 400 million. Integration of the methodology in the formal policy or in the form of web-based decision support tool may reach these people effectively. Need not to say, the same may be explored anywhere in the world using flexible and participatory indicator selection and their subsequent validation.

AUTHORS' CONTRIBUTIONS

RG designed research, analysed and interpreted data and prepared the manuscript, SC conceptualized and designed the research work, and BP collected data and drafted the manuscript with RG. All authors read and approved the final manuscript.

ACKNOWLEDGEMENTS

Logistic supports received from the All India Coordinated Research Project on Integrated Farming System and Academic Research Programme of Ramakrishna Mission Vivekananda University is duly acknowledged. Authors are also thankful to all the farmers who assisted and participated in the survey.

REFERENCES

1. Abdi H (2007) Discriminant Correspondence Analysis. In: Salkind NJ (ed) Encyclopedia of Measurement and Statistics, Sage, Thousand Oaks, CA.
2. Aggarwal PK, Hebbar KB, Venugopalan MV, Rani S, Bala A, Biswal A, Wani SP (2008) Quantification of yield gaps in rainfed rice, wheat, cotton and mustard in india. ICRISAT, Hyderabad.
3. Anderberg MR (1973) Cluster Analysis for Applications. Academic, New York.
4. Andersen E (2009) Regional Typologies Of Farming Systems Contexts, seamless Integrated Project, Eu 6th Framework Programme.
5. Andersen E, Elbersen B, Godeschalk F, Verhoog D (2009) Farm management indicators and farm typologies as a basis for assessments in a changing policy environment. J Env Manage 82:353-362
6. Bagchi BD, Roy SB, Jaim WMH, Hossain M (2012) Diversity, spatial distribution, and the process of adoption of improved rice varieties in West Bengal. In: Hossain M, Jaim WMH, Paris TR, Hardy B (eds) Adoption and diffusion of modern rice varieties in Bangladesh and eastern India, IRRI, Philippines. pp 31-44
7. Bahu RR, Mishra P (2001) Farm ponds in Coastal Saline Zone of West Bengal: a techno-economic feasibility study. J Agric Eng 38:17-22
8. Basu D, Banerjee S, Goswami R (2009) Vegetable cultivation on ail (bund): a successful farmer-led technology transfer. Acta Hort 809:155-160

9. Bhattacharyya R (2008) Crop Diversification: A Search For An Alternative Income Of The Farmers In The State Of West Bengal In India.
10. Bhattacharjee S, Saravanan R (2012) Poverty In Natural Prosperity: Can Agriculture Bring The Renaissance In North-East India? In: Kashyap SK, Pathak A, Papnai G (eds) Saving Humanity Saving Humanity, Swami Vivekananda Perspective. Vivekanand Swadhyay Mandal, GBPUAT, Pantnagar, India. pp 260-274
11. Bidogeza JC, Berentsen PBM, de Graaff J, Lansink AGJMO (2009) A typology of farm households for the Umutara Province in Rwanda. Food Sec 1:321-335
12. Birner R, Anderson JR (2007) How To Make Agricultural Extension Demand-Driven? The case of india's agricultural extension. International Food Policy Research Institute, Washington DC.
13. Birner R, Davis K, Pender J, Nkonya E, Anandajayasekeram P, Ekboir J, Mbabu A, Spielman D, Horna D, Benin S, Kisamba-Mugerwa W (2009) From best practice to best fit: a framework for designing and analyzing pluralistic agricultural advisory services worldwide. J Agric Edn Ext 15:341-355
14. Bozeman B (2000) Technology transfer and public policy: a review of research and theory. Res Policy 29:627-655
15. Briggeman BC, Gray AW, Morehart MJ, Baker TG, Wilson CA (2007) A new U.S. farm household typology: Implications for agricultural policy. Rev Agric Econ 29:765-782
16. Chambers R, Pacey A, Thrupp LA (1989) Farmer First: Farmer Innovation And Agricultural Research. Intermediate Technology Publications, London.
17. Conroy C, Sutherland A (2004) Participatory technology development with resource-poor farmers: maximising impact through the use of recommendation domains. ODI, London.
18. De UK (2002) Economics Of Crop Diversification. Akansha Pub, House, New Delhi.
19. De UK (2003) Diversification of crop in West Bengal: a spatio-temporal analysis. Artha Vijnana 42:170-182
20. Deb UK, Rao GN, Rao YM, Slater R (2002) Diversification And Livelihood Options: A Study Of Two Villages In Andhra Pradesh, India. Department for International Development (DFID), United Kingdom.

21. Dhiman SC, Thambi DS (2009) Ground water management in coastal areas. Bhu-Jal 24:69-73
22. Ding C, He X (2004) K-Means Clustering Via Principal Component Analysis. In: Proceedings Of The Twenty-First International Conference On Machine Learning. , Banff, Canada. [http://ranger.uta.edu/~chqding/papers/KmeansPCA1.pdf] .Available via http://ranger.uta.edu/~chqding/papers/KmeansPCA1.pdf. Accessed 16 Aug 2013
23. Dixon J, Gulliver A, Gibbon D (2001) Farming Systems and Poverty: Improving Farmers' Livelihoods In A Changing World. FAO, Rome.
24. Ellis F (1993) Peasant Economics: Farm Households and Agrarian Development. Cambridge University Press, Cambridge.
25. Emtage N, Suh J (2005) Variations in socioeconomic characteristics, farming assets and livelihood systems of Leyte rural households. Ann Trop Res 27:35-54
26. Engel PGH (1995) Facilitating Innovation: An Action-Oriented Approach and Participatory Methodology To Improve Innovative Social Practice In Agriculture. Wageningen Agricultural University, Dissertation.
27. Fafchamps M (1992) Cash crop production, food price volatility, and rural market integration in the third world. Am J Agric Econ 74:90-99
28. Fernandez-Cornejo J, Hendricks C, Mishra A (2005) Technology adoption and off-farm household income: the case of herbicide-tolerant soybeans. J Agric Appl Econ 37:549-63
29. Field A (2005) Discovering Statistics Using Spss For Windows. Sage, London.
30. Fliegel F, Kivlin JE, Sekon GS (1968) A cross-national comparison of farmers' perceptions of innovations as related to adoption behavior. Rural Sociol 33:437-449
31. Folmer H, Dutta S, Oud H (2010) Determinants of rural industrial entrepreneurship of farmers in West Bengal: a structural equations approach. Int Region Sc Rev 33(4):367-396
32. Frankenberger TR, Dewalt BR, McArthur HJ, Hudgens RE, Mitawa G, Rerkasem K, Finan T, Butler Flora C, Young N (1989)

Identification of results of farming systems research and extension activities: a synthesis. FSRE Newsl 1:8-10

33. Gajbhiye KS, Mandal C (2008) Agro–Ecological Zones, Their Soil Resource and Cropping Systems.
34. Ganpat W, Bekele I (2001) Looking For The Trees In The Forest: Farm Typology As A Useful Tool In Defining Targets For Extension. In: Lindner JR (ed) Proceedings of the 17th Annual Conference of the Association for International Agricultural and Extension Education, , Baton Rouge, Louisiana. [http:/ / agris.fao.org/ agris-search/ search.do?f=2002/ TT/ TT02001.xml;TT2002001006] Available via http://agris.fao.org/agris-search/search.do?f=2002/TT/TT02001.xml;TT2002001006. Accessed 22 February 2014
35. Gebauer RH (1987) Socio-economic classification of farm household - conceptual, methodical and empirical considerations. Eur Rev Agric Econ 14:261-283
36. Ghosh BK, Kuri PK (2005) Changes in cropping pattern in West Bengal during 1970-71 to 2000-2001. IASSI Quart 24:39-56
37. Glendenning CJ, Babu S, Asenso-Okyere K (2010) Review Of Agricultural Extension In India. Are Farmers Information Needs Being Met?. IFPRI, Washington DC.
38. Gómez-Limón JA, Berbel J, Arriaza M (2007) MCDM Farm System Analysis For Public Management Of Irrigated Agriculture. In: Weintraub A, Bjorndal T, Epstein R, Romero YC (eds) Handbook On Operations Research In Natural Resources In International Series In Operations Research And Management Science, Páginas, Kluwer, New York. pp 93-114
39. Gorsuch RL (1983) Factor Analysis. Erlbaum, Hillsdale, NJ.
40. Goswami R (2007) Understanding Farmer-To-Farmer Communication Within The Sustainable Rural Livelihood Framework. Dissertation, Bidhan Chandra Krishi Viswavidyalaya.
41. Goswami R, Biswas MS, Basu D (2012) Validation of participatory farming situation identification: a case of rainfed rice cultivation in selected area of West Bengal, India. Ind J Tradition Knowl 11:471-479
42. Go WB (2005) District Statistical Handbook. Bureau of Applied Economics & Statistics, Government of West Bengal, India.

43. Guto SN, Pypers P, Vanlauwe B, de Ridder N, Giller KE (2010) Socio-ecological niches for minimum tillage and crop-residue retention in continuous maize cropping systems in smallholder farms of Central Kenya. Agron J 104:188-198

44. Halder JC, Das P (2012) Present status and futuristic view of horticulture in West Bengal. Geo-Analyst 2(1):[http://gswb.in/wp-content/uploads/2012/08/v2n1jully2012_1.pdf] Available via http://gswb.in/wp-content/uploads/2012/08/v2n1jully2012_1.pdf. Accessed 10 May 2013

45. Hall AJ, Yoganand B (2004) New Institutional Arrangements In Agricultural Research And Development In Africa: Concepts And Case Studies. In: Hall AJ, Sulaiman RV, Raina RS, Shambu Prasad C, Naik GC, Clark NG (eds) Innovations In Innovation: Reflections On Partnership, Institutions And Learning, International Crops Research Institute for the Semi-Arid Tropics/National Centre for Agricultural Economics and Policy Research, Andhra Pradesh/New Delhi. pp 105-134

46. Hardiman RT, Lacey R, Yang MY (1990) Use of cluster analysis for identification and classification of farming systems in Qingyang County, Central North China. Agric Syst 33:115-125

47. Harrington LH, Tripp R (1984) Recommendation domains: a framework for on-farm research. CIMMYT Economics Working Paper 02/84. Series: CIMMYT Economics Working Paper, CIMMYT, Mexico.

48. Harris RJ (2001) A Primer Of Multivariate Statistics. Academic, New York.

49. Hayami Y, Ruttan VW (1981) Agricultural Development: An International Perspective. Johns Hopkins University Press, Baltimore, USA.

50. Hazell PB (2005) Is there a future for small farms? Agric Econ 32:93-101

51. Jolliffe IT (2002) Principal Component Analysis. Springer Verlag, New York.

52. Jolly CM (1988) The use of action variables in determining recommendation domains: senegalese farmers for research and. Agric Admin Ext 30:253-267

53. Kaiser HF (1970) A second generation Little-Jiffy. Psychomelrika 35:401-415
54. Kaufman L, Rousseeuw PJ (1990) Finding groups in data: an introduction to cluster analysis. Wiley, New York.
55. Khanna SS (1989) The Agro-Climatic Approach. In: Survey of Indian Agriculture. The Hindu, Madras, India. pp 28-35
56. Kherallah M, Kirsten JF (2002) The New Institutional Economics: applications for agricultural policy research in developing countries. Agrekon 41:111-134
57. Kobrich C, Rehman T, Khan M (2003) Typification of farming systems for constructing representative farm models: two illustrations of the application of multi-variate analyses in Chile and Pakistan. Agric Syst 76:141-157
58. Kostrowicki J (1977) Agricultural typology concept and method. Agric Syst 2:33-45
59. Kotler P (1986) Principles of Marketing. Prentice-Hall, Englewood Cliffs, NJ.
60. Kydd J (2002) Agriculture And Rural Livelihoods: Is Globalisation Opening Or Blocking Paths Out Of Rural Poverty?. Overseas Development Institute (ODI), London, UK.
61. Lal P, Lim-Applegate H, Scoccimarro M (2001) The adaptive decision making process as a tool for integrated natural resource management: focus, attitudes and approach. Conserv Ecol 5:11
62. Leeson JY, Sheard JW, Thomas AG (1999) Multivariate classification of farming systems for use in integrated pest management studies. Can J Plant Sci 79:647-654
63. Mahapatra AK, Mitchell CP (2001) Classifying tree planters and non planters in a subsistence farming system using a discriminant analytical approach. Agroforest Syst 52:41-52
64. Malik DP, Singh IJ (2002) Crop diversification – an economic analysis. Ind J Agric Res 36:61-64
65. Mehta R (2009) Rural Livelihood Diversification and Its Measurement Issues: Focus India. Wye City Group on Statistics On Rural Development And Agriculture Household Income, Second Meeting, at FAO HQ, Italy, Rome, 11-12 June 2009.
66. Mohammad A (1978) Situation Of Agriculture, Food, And Nutrition In Rural India. Concept, New Delhi.

67. Mwijage NA, de Ridder N, Baijukya F, Pacini C, Giller EK (2009) Exploring the variability among smallholder farms in the banana-based farming systems in Bukoba district, Northwest Tanzania. Afr J Agric Res 4:1410-1426

68. (2009) State Agricultural Plan of Government of West Bengal. NABARD Consultancy Service Private Limited. , Kolkata, India.

69. Nabi R (2008) Constraints to the adoption of rice-fish farming by smallholders in Bangladesh: a farming systems analysis. Aquacult Econ Manage 12:145-153

70. Nehring R, Fernandez-Cornejo J, Banker D (2005) Off-farm labour and the structure of U.S. Agriculture: the case of Corn/ Soybean farms. Appl Econ 10:633-650

71. North DC (1997) Prologue In The Frontiers Of New Institutional Economics. In: Drobak JN, Nye VCJ (eds) The Frontiers of The New Institutional Economics, Academic, New York.

72. Ojiem J, Ridder N, Vanlauwe B, Giller KE (2006) Socio-ecological niche: a conceptual framework for integration of legumes in smallholder farming systems. Int J Agric Sust 4:79-93

73. Orr A, Jere P (1999) Identifying smallholder target groups for IPM in southern Malawi. Int JPest Manage 45:179-187

74. Pal S, Mruthyunjaya J, Joshi PK, Saxena R (2003) Institutional change in Indian agriculture. National Centre for Agricultural Economics and Policy Research. Indian Council of Agricultural Research, New Delhi.

75. Pender J, Hazell P (2001) Promoting Sustainable Development In Less-Favored Areas, IFPRI Policy Brief, Focus 4. IFPRI, New York.

76. Prein M (2002) Integration of aquaculture into crop–animal systems in Asia. Agric Syst 71:127-146

77. Radulovic V (2005) Are new institutional economics enough? Promoting photovoltaics in India's agricultural sector. En Policy 33:1883-1899

78. Reckling M (2011) Characterisation of Bean Farming System Across Farm Types in Northern and Eastern Rwanda.: Identification Of Potential Niches For Legume Technologies. Dissertation, Wageningen University, The Netherlands.

79. Rogers EM (1962) Diffusion of Innovations. The Free Press, New York.

80. Rogers EM (1995) Diffusion of innovation. The Free Press, New York.
81. Roling N (1988) Extension Science: Information Systems In Agricultural Development. Cambridge University Press, Cambridge.
82. Rowe EC, van Wijk MT, de Ridder N, Giller KE (2006) Nutrient allocation strategies across a simplified heterogeneous African smallholder farm. Agric Ecosyst Environ 116:60-71
83. Ruttan VW, Hayami Y (1984) Toward a theory of induced institutional innovation. J Dev Stud 20:203-223
84. Sanjeev MV, Venkatasubramanian V, Singha AK (2010) Farming Dystems Of North East India: Research And Development Strategies for KVKs. Zonal Project Directorate, Zone – III. Indian Council of Agricultural Research, Umiam, Meghalaya, India.
85. Savadogo K, Reardon T, Pietola K (1998) Adoption of improved land-use technologies to increase food security in Burkina Faso: relating animal traction, productivity, and nonfarm income. Agric Syst 58:441-464
86. Senthilkumar K, Bindraban PS, de Boer W, de Ridder N, Thiyagarajan TM, Giller KE (2009) Characterising rice-based farming systems to identify opportunities for adopting water efficient cultivation methods in Tamil Nadu, India. Agric Water Manage 96:1851-1860
87. Singh S (2002) Contracting out solutions: political economy of contract farming in Indian Punjab. World Dev 30:1621-1638
88. Singh AK, Singh S, Venkatasubramanian V (1993) Relationship between farmers. Socio- personal traits and adoption of dairy innovations. Charion 22:84-87
89. Soule MJ (2001) Soil management and the farm typology: do small family farms manage soil and nutrient resources differently than large family farms? Agric Resour Econ Rev 30:179-188
90. Sukhatme PV, Sukhatme BV, Sukhatme S, Asok C (1984) Theory Of Sample Surveys With Applications. ISAS, IASRI, New Delhi.
91. (2006) Trade Reforms And Food Security: Country Case Studies And Synthesis. Food and Agriculture Organisation of UN, Rome.

92. Timothy WO (1994) Identifying target groups for livestock improvement research: the classification of sedentary livestock producers in western Niger. Agric Syst 46:227-237

93. Tittonell P, Vanlauwe B, Leffelaar PA, Shepherd KD, Giller KE (2005) Exploring diversity in soil fertility management of smallholder farms in western Kenya. II. Within-farm variability in resource allocation, nutrient flows and soil fertility status. Agric Ecosyst Environ 110:166-184

94. Tittonell P, Leffelaar PA, Vanlauwe B, van Wijk MT, Giller KE (2006) Exploring diversity of crop and soil management within smallholder African farms: a dynamic model for simulation of N balances and use efficiencies at field scale. Agric Syst 91:71-101

95. Tittonell P, Vanlauwe B, de Ridder N, Giller KE (2007) Heterogeneity of crop productivity and resource use efficiency within smallholder Kenyan farms: soil fertility gradients or management intensity gradients. Agric Syst 94:376-390

96. Tittonell P, Muriukid A, Shepherde KD, Mugendif D, Kaizzig KC, Okeyoa J, Verchote L, Coee R, Vanlauwea B (2010) The diversity of rural livelihoods and their influence on soil fertility in agricultural systems of East Africa – A typology of smallholder farms. Agric Syst 103:83-97

97. Usai MG, Casu S, Molle G, Decandia M, Ligios S, Carta A (2006) Using cluster analysis to characterize the goat farming system in Sardinia. Livest Sci 104:63-76

98. (2000) ERS Farm Typology For A Diverse Agricultural Sector. USDA, US.

99. Vanclay JK (2005) Using a typology of tree-growers to guide forestry extension. Ann Trop Res 27:97-103

100. Williamson OE (1990) A comparison of alternative approaches to economic organization. J InstTheor Econ 146:61-71

101. Zingore S, Murwiraa HK, Delvea RJ, Gillerb KE (2007) Soil type, management history and current resource allocation: Three dimensions regulating variability in crop productivity on African smallholder farms. Field Crops Res 101(3):296-305

Chapter 6

Caught in a Sandstorm: An Assessment of Pressures on Communal Pastoral Livelihoods in the Kalahari Desert of Botswana

Diphetogo Mosalagae[1] and Kabo Mogotsi[2]

[1]Animal Production and Range Research Division, Department of Agricultural Research, Ministry of Agriculture, Private Bag 0033, Sebele, Botswana

[2]Animal Production and Range Research Division, Department of Agricultural Research, Ministry of Agriculture,

ABSTRACT

Despite their importance to rural livelihoods, the economic and environmental sustainability of contemporary communal pastoral systems of the Kalahari is increasingly being put under scrutiny. Using data collected from a survey of randomly selected pastoral households from two districts of Ghanzi and Kgalagadi South, Botswana, this paper outlines the major constraints affecting contemporary extensive pastoral systems. Because of the prevailing semi-arid climate, recurrent drought outbreaks were cited as a common challenge in both study areas together with lack of fenced grazing areas. Stock theft was also highlighted as a challenge, even though it was more pronounced in Kgalagadi South. In Ghanzi area, the other important challenges included loss of livestock due to predation, while stray animals, stock theft and loss of livestock due to road accidents were equally important. Kgalagadi South pastoral households experienced serious shortage of drinking water for their livestock, followed by inadequate grazing land and limited access to distant markets. The preceding, together with several other less prominent challenges revealed in this study, exert pressure on pastoral livelihoods. There is need, therefore, to address these perceived challenges and strengthen resilience through effective policy reviews to secure rural livelihoods.

BACKGROUND

The Kalahari desert ecosystem of Botswana is characterized by inherent climatic uncertainty, particularly rainfall variability, and consequently exhibits low and highly variable biomass productivity. As a result, wildlife and extensive pastoral-based livelihoods have evolved over time as an adaptation to this 'marginal' environment, and fewer alternative land use systems, if any, can claim to be more economically efficient and environmentally sustainable. Livestock production is central to the livelihood of the majority of rural communities in semi-arid Botswana, and it is characterized by two distinct land tenure systems, communal and commercial. Apart from providing subsistence needs such as meat, milk and hides, livestock - particularly cattle - are also socio-economically important (Keijsper 1993; Mrema and Rannobe 1996). The latter use of livestock, however, has often been undervalued (Abel 1997;Behnke 2008) or simplified due to the growing commercialization of beef (Darkoh and Mbaiwa 2003). Despite the apparent value of livestock to the rural economy, the agricultural sector has experienced unprecedented decline in overall performance over the years. The sector's contribution to the national economy Gross Domestic Product (GDP) has declined from 40% since 1966 to less than 2% currently (GoB 2009a), mainly attributed to the growing importance of mining and tourism sectors. The beef sub-sector still contributes significantly to the total agricultural GDP.

To better understand the context of pastoral changes in Botswana, it is necessary to highlight past major policy shifts. Under the traditional system of land tenure, land was always regarded as communal and tribal chiefs allocated land for arable and/or grazing use. The traditional system was replaced in 1968 by the Tribal Land Act which introduced decentralized Land Boards to administer land, instead of chiefs. The passing of the Tribal Land Act and the establishment of Land Boards did not fundamentally change the concepts of land tenure (Greenhow 1978). However with time, the traditional communally-owned rangelands in the eastern Kalahari were perceived by the government to be overstocked and overgrazed by livestock and thus degraded. Buying into the 'tragedy of the commons' narrative popularized by Hardin (1968), the Tribal Grazing Land Policy (TGLP) advocated for structural reforms to the country's livestock industry, to ultimately enhance national economic and social development (GoB 1975). Subsequently, part of the tribal grazing land was demarcated into ranches. The specific objectives of TGLP were to:

1. Improve range management by preventing overgrazing and further degradation through paddocking and rotational grazing. Individual owners or syndicates of large herds (>400 animals) were to be moved off communal lands into fenced, borehole-focused ranches of uniform size.
2. Increase livestock productivity and farmers' income by promoting better management practices such as daily watering, selective breeding and early weaning.
3. Promote social equity by retaining only small-scale farmers on communal lands and thus allowing some room to improve their livelihoods.
4. Secure the interest of future generations and/or those who were not livestock farmers by reserving some land.

To ease the environmental and social conundrum in the wetter eastern Botswana, which held 70% of the national herd (Thomas and Sporton 1997), the TGLP encouraged development further west into the more arid Kalahari desert, aided by borehole installations. But with time, it became apparent that the TGLP was based on false assumptions. Firstly, there was not enough land to demarcate into viable livestock ranches as previously assumed. Secondly, overriding abiotic factors like rainfall, fire and drought characteristic of the non-equilibrium nature of semi-arid and arid environments called into question the coupling of plant-livestock relations emphasized by the equilibrium theory in grazing systems (Behnke et al. 1993; Scoones 1995). Since density of herbivores is not the only driver of vegetation dynamics, it followed that destocking per se as a management tool would not be effective. The final assumption of TGLP was that those allocated ranches will stay as ranches. To the contrary, ranch owners employed dual grazing rights, where they let their herds onto the commons during wet grazing periods and retreated to their exclusive ranches during dry seasons. Overall, extensive reviews of the TGLP have concluded that it failed to address its objectives (Tsimako 1991; White 1993; Frimpong 1995).

Upon realizing the shortcomings of TGLP, the government introduced the National Policy on Agricultural Development (NPAD) in 1991. The aim of NPAD was to provide subsidies to improve productivity in the livestock sub-sector by promoting the sustainable use of rangeland resources through fencing of communal grazing areas (GoB 1991). So in essence, the fencing component of NPAD simply accelerated the implementation of TGLP and failed to tackle some of the earlier concerns like dual grazing rights. But not everyone was practicing pastoralism. Some ranches had displaced non-pastoral populations and denied them access to natural resources of land, water, veld products and animals (Hitchcock 1978). So for other traditional rural livelihoods such as hunting and gathering, part of the reserve land was later demarcated into Wildlife Management Areas (WMAs) in an attempt to balance the scales. WMAs are multiple-use areas combining wildlife conservation with the creation of economic opportunities for the rural population (GoB 1999). Intense developments within WMAs such as borehole drilling are prohibited, while hunting is controlled through permits or total bans. There are complementary polices and pieces of legislation predicated on sustainable utilization of natural resources also governing these areas, such as the Wildlife Conservation Policy (1986), Tourism Policy (1990), National Conservation Strategy (1990), Wildlife Conservation and National Parks Act (1992), Tourism Act (1992) and Community Based Natural Resource Management (CBNRM) concept.

Given the historical context above, the current paper focuses on contemporary pastoral livelihoods under communal land tenure. While once viewed as destructive, economically irrational and archaic (GoB 1975; FAO 2001), the extensive livestock production system continues to directly and/or indirectly support a large rural population and the importance of livestock still transcends many sectors. Approximately 39% - 791,700 people - of Botswana's human population resides in rural areas, out of a total of 2.03 million. Botswana's cattle population stood at 2.55 million in 2011 (GoB 2013), where grazing in communal rangelands accounted for 86% of the cattle and 71% of farmers in Botswana, while private grazing in ranches accounted for 14% of the national cattle herd and 5% of the land area. Though the traditional communal, open access grazing lands are often dismissed as unproductive relative to modern ranches (APRU 1980), empirical evidence does not always reflect this (Behnke 1985; deRidder and Wagenaar 1986). For example, looking at previous work in Botswana, Davies and Hatfield (2007) noted that production in cash, energy and protein terms per hectare under communal areas exceeds by at least three times per hectare returns from ranches, even though technical production parameters are lower. The latest statistics show that the traditional sector has consistently had higher cattle birth rates than the commercial sector (54.4% and 38.9%, respectively), but this achievement is compromised by the high mortality rates observed in communal lands relative to commercial ranches (GoB 2013). The rate of mortality in communal grazing lands was 6.6%, while commercial ranches experienced 1.6% mortality rate. Off-take rates in the commercial sector increased from 12.5% in 2010 to 13.5% in 2011 while the national rate dropped from 8.0% to 6.9% during the same period (GoB 2013).

If the livestock sector is to continue being the mainstay of rural livelihoods in future, then there should be deliberate efforts to maintain and/or improve its productivity. But first, the factors linked to the observed decline in livestock productivity need to be identified, understood and appreciated. While some factors have been perceived as affecting livestock production in Botswana (GoB 1991; Mrema and Rannobe 1996; Burgess 2002; CAR 2005; BIDPA 2006), these are more often than not at national level and thus grossly aggregated, with a few exceptions (e.g. Kgosikoma 2006). Furthermore, the process of identifying problems at the local level is influenced by 'outsiders' in the form of environmental managers/experts and policy-makers and may end up not receiving support from the community (Fraser et al. 2006). A more comprehensive approach must consider the interests, positions and needs of pastoralists so as to guide decision-makers and develop a suitable and legitimate process of communal land and resource tenure that fits with both the priorities of pastoralists as well as the government (Flintan and Cullis 2010).

This study therefore set out to identify and characterize the major constraints faced by contemporary extensive livestock production systems in the west and south-western semi-arid Kalahari of Botswana, as perceived by the pastoral communities. Knowledge about challenges hampering the livestock sector will guide the timing and form of intervention measures, as well as better inform policy designed to halt and reverse the decline in livestock production and hence ensure continued sustainability of rural livelihoods.

Study Area

The study was conducted in the western Kalahari region of Botswana, in Ghanzi and Kgalagadi Districts. The Ghanzi study area covered Ghanzi village (21°42′S, 21°49′E) and the surrounding communal grazing lands. The area is characterized by deep, infertile Kalahari sands. Rainfall is unimodal and averages 375 mm per annum. Maximum mean daily temperatures reach 33°C to 45°C and 22°C in summer and winter months, respectively, with a minimum of 4°C to −5°C in winter months. The vegetation is classified as tree and bush savanna, dominated by woody species such as Terminalia sericea and Boscia albitrunca, while the herbaceous layer includes grasses likeEragrostis lehmanniana,Pogonarthria squarrosa and Digitaria eriantha.

The second study area was Kgalagadi South sub-district within the Kgalagadi District and covered Tsabong village (26°112″S, 22°24′20″E) and the surrounding communal grazing areas. Rainfall also follows a unimodal pattern, falling mainly between the months of November and March with a long-term annual average of less than 250 mm and high variability. Maximum temperatures reach 45°C in summer, and minimum winter temperatures can be as low as −4°C. The vegetation is classified as shrub/bush savanna, dominated by woody species such as Acacia erioloba, Acacia mellifera and Acacia luederitzii, while grasses include E.lehmanniana, Stipagrostis uniplumis andSchmidtia kalahariensis.

The introduced exotic Prosopis tree species has increasingly become an important invader, especially in proximity to human settlements.

Towards the extreme south-west, bare rolling sand dunes are visible and in some areas are covered by layers of grass.

In both study areas, small-scale agriculture is practiced, mainly dominated by livestock rearing in communal, open access grazing lands referred to as cattle posts or locally known as meraka. Beef and game ranching is also notably present in this part of the country. Limited small-scale rain-fed crop cultivation is spread over both the districts, mainly including maize, sorghum, beans and melons. Though both study areas lie within the Kgalagadi (Kalahari) desert ecosystem, the second study site of Kgalagadi South is in the driest and least inhabited region of Botswana, and thus, in some areas crop cultivation is not practiced at all. This widened geographical scope enabled the study to capture differences that may be due to climatic and related ecological differences.

METHODS

Data Collection

In each purposively selected study area, pastoral households[a] were randomly selected with the aid of extension agents from the Ministry of Agriculture and the local authorities. Because absentee ownership is common, appointments were made in advance with heads of each household, who were then interviewed at their preferred time. In Ghanzi area, 24 pastoral households were surveyed,while 16 were chosen for the Kgalagadi South study area. The Kgalagadi South area, as earlier noted, is the driest and least inhabited in Botswana. Subsequently, pastoral households are relatively fewer and sparsely distributed, thus resulting in a serious challenge in terms of accessibility through the sandy terrain. The size of the area to be covered as well as the very nature of pastoral systems located in remote areas meant that only four-wheel-drive vehicles could be used.

[a]Following Elhadi et al. (2012), a household refers to all people who live under one roof and are subject to decisions made by the household head. The latter is the owner of the main source of income or means of production in a household and therefore provides the basic needs for the household members.

A structured questionnaire with open-ended, multiple-response and dichotomous questions was used to collect data. A pilot test run was conducted prior to the main survey, and the final questions were amended accordingly. The survey questionnaire collected data on socio-economic variables of households and their livestock management practices including feed and water resources, markets, labour usage as well as their perceived constraints hampering the sustainability of the extensive livestock production system under investigation. To avoid misinterpretation, the interviews were conducted in Setswana, the national language generally understood by the respondents. In addition to direct observations in the field (including participant observation), conversational exchanges with key informants such as extension agents, representatives of farmer groups/committees and local headmen/chiefs complemented the survey.

Data Analysis

Data were coded and thereafter analyzed using the frequency procedure of SAS (SAS 2005) for socio-economic characteristics of respondents and related variables. The chi-square (χ^2) test was used to confirm whether respondents were uniformly distributed or not among the groups for each variable or, for the two-way tables, to measure the interaction or strength of association between any two variables. The inferential tests were declared significant or not using the calculated χ^2 testp value against the 5% level of significance.

RESULTS

Socio-Economic Characteristics

The socio-economic characteristics of the respondents are shown in Table 1.

Table 1: The socio-economic characteristics of pastoral households in the study areas

Variables	Study area		
	Ghanzi (%)	**Kgalagadi South (%)**	**Total (%)**
Gender[a]			
(i) Male	19 (48)	15 (38)	34 (85)
(ii) Female	5 (13)	1 (3)	6 (15)
Age group[a]			
(i) 21 to 39 years	4 (10)	2 (5)	6 (15)
(ii) 40 to 65 years	19 (48)	8 (20)	27 (63)
(iii) >65 years	1 (3)	6 (15)	7 (18)
Educationlevel[a]			
(i) Illiterate	7 (18)	5 (13)	12 (30)
(ii) Formal education	17 (43)	11 (28)	28 (70)
Main source of income			
(i) Employed in agriculture	13 (33)	14 (35)	27 (68)
(ii) Formal employment	7 (18)	2 (2)	9 (23)
(iii) Pension	4 (10)	0 (0)	4 (10)
Livestock ownership			
(i) Cattle	24 (65)	13 (35)	37 (100)
(ii) Goats	16 (52)	15 (48)	31 (100)
(iii) Sheep	5 (31)	11 (69)	16 (100)
(iv) Equine	19 (54)	16 (46)	35 (100)
(v) Chickens	19 (95)	1 (5)	20 (100)

[a]Head of household.

Mosalagae and Mogotsi

Mosalagae and Mogotsi Pastoralism: Research, Policy and Practice 2013 3:18 doi:10.1186/2041-7136-3-18

The pastoral households surveyed were mostly headed by older males (≥40 years) in both study areas, and the majority (70%) had some formal education. In Ghanzi, the main sources of income in decreasing order of importance were as follows: sales from agricultural activities, formal employment and monthly pension. Kgalagadi South households earned their income mainly from animal and crop agriculture, and only 2% had some form of formal employment. The most reared and highly valued livestock were cattle and small stock (goats and sheep), though cattle dominated in Ghanzi while more households kept small stock in

Kgalagadi South area. Equines, mostly horses and donkeys, were also reared in relatively high numbers, commonly used as transport when traversing the extensive deep sand and predator-filled Kalahari terrain.

Livestock Management Practices

The main livestock management practices undertaken by pastoral households in the study areas are shown in Table 2.

Table 2: Some livestock management practices undertaken in the study areas

Variables	Study area		
	Ghanzi (%)	**Kgalagadi South (%)**	**Total (%)**
Source of livestock drinking water			
(i) Own borehole	6 (15)	7 (18)	13 (33)
(ii) Syndicate borehole	18 (45)	9 (23)	27 (68)
Gender of labour			
(i) Males	11 (28)	15 (38)	26 (65)
(ii) Females	2 (5)	0 (0)	2 (5)
(iii) Males and females	11 (28)	1 (3)	12 (30)
Reliability of labour			
(i) Very reliable	16 (40)	3 (8)	19 (48)
(ii) Reliable	5 (13)	7 (18)	12 (30)
(iii) Not reliable	3 (8)	5 (13)	8 (20)
(iv) Do not know	0 (0)	1 (3)	1 (3)
Type of livestock housing			
(i) Bush enclosures (kraals)	1 (3)	2 (5)	3 (8)
(ii) Improved shelter	23 (55)	14 (35)	37 (93)
Mineral supplementation of livestock?			
(i) Yes	22 (55)	15 (38)	37 (93)
(ii) No	2 (5)	1 (3)	3 (8)

Mosalagae and Mogotsi

Mosalagae and Mogotsi Pastoralism: Research, Policy and Practice 2013 **3**:18 doi:10.1186/2041-7136-3-18

The Kalahari desert is characterized by the absence of surface water for most periods of the year. Thus, in this study, most pastoral households relied on underground water through sinking of boreholes. Syndicate boreholes were used more (68%) in both study areas, where a group of pastoral households shared the same water point. Syndicates are organizations in which a group of people come together for the common purpose of owning and operating a borehole (Hitchcock1978), and members contribute equally towards the cost of drilling, daily running and maintenance of the borehole. The distances from the homesteads to livestock watering points varied among households. In Ghanzi, 71% of households indicated that their main source of drinking water for livestock was within a distance of 1km, while 29% had to travel for more than 1 km to livestock water points. A similar trend was observed in Kgalagadi South, where 75% and 25% of households had to cover distances of less than 1 km and greater than 1 km, respectively, to reach livestock water points.

The livestock were housed under improved structures (93%), which were mostly built using treated poles and fence, while some even had feed and water troughs as well as shaded areas from corrugated iron sheets. The remainder of the households built simple bush enclosures using branches of the thorny Acacia tree species. The animals were housed in one group, except where young calves were separated from the rest of the herd overnight to allow hand milking of lactating cows in the morning for household consumption. Pregnant cows, lactating cows and calves grazed closer to the homesteads as they require water more often than other types of livestock and also for safety from marauding predators.

In contrast to most traditional extensive pastoral systems (Coppock and Sovani 1999; FAO 2001; ESAP 2003), there was notable mineral supplementation of livestock undertaken in both study areas even though the natural rangeland still provided the bulk of the livestock feed. Apart from the popularly used coarse salt (88%) and dicalcium phosphate (63%), other supplements included beef finisher meal (30%), bull ration (15%), drought pellets (13%) and assorted minor supplements of sorghum bran/molasses/mineral blocks (40%). The use of these supplements, however, was mostly done on an ad hoc basis, and it was more pronounced during the dry season.

The gender of workers in surveyed pastoral systems was highly skewed towards males (65%). Labour is particularly essential for herding and in filling troughs with drinking water for cattle. Herding varies with seasons and may cover distances of up to 20 km daily. In addition, other duties include everyday livestock handling activities such as milking, branding, dehorning, castration, hoof trimming and occasional maintenance of borehole and livestock housing structures. In essence, labourers are expected to multi-task and be flexible. Only a relatively small proportion (20%) of labourers was deemed as unreliable. While labourers can plan and execute daily chores on their own, the 'serious' decision-making powers are still exclusively retained by the livestock owner. These decisions include, among others, selling or killing of any livestock, granting approval for new incoming herds on the same grazing area, as well as when and where to move animals in times of extended drought periods.

Production Constraints

The challenges faced by pastoral communities in Ghanzi and Kgalagadi South are reflected in Figure 1.

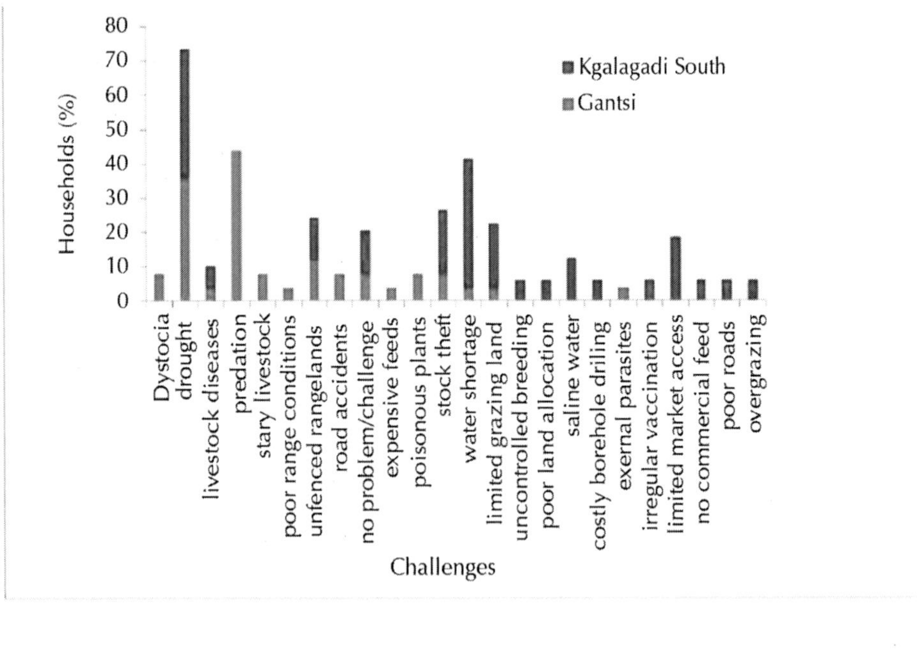

Figure 1: Challenges faced by pastoral households in Kgalagadi South and Ghanzi study areas.

In Ghanzi area, the most important challenges included loss of livestock due to predation (44%), drought (36%) and lack of fenced grazing areas (12%), while stray animals, stock theft and loss of livestock due to road accidents were all equally important (8% each). Kgalagadi South pastoral households experienced problems of drought and shortage of drinking water for livestock (38% each), followed by stock theft, limited grazing land, limited access to distant markets (19% each) and lack of fenced grazing areas (13%). As many as 23 constraints were encountered by pastoralists overall, but only 6 were common to both study areas. Though numerous, some of the mentioned challenges were complex and intricately linked. For example, lack of fenced grazing areas, uncontrolled breeding, stray animals and predation by carnivores are all functionally linked. Another example of associated challenges is overgrazing, poor range conditions, diseases and poisonous or injurious plants killing livestock. Only 8% and 13% of households did not mention any challenges in Ghanzi and Kgalagadi South, respectively.

DISCUSSION

Production Constraints

Only the five most important constraints, as perceived by pastoral households, are further discussed next. Both areas experienced drought. In Ghanzi, other constraints included predation, drought, unfenced grazing areas, stray animals and loss of livestock to road accidents. Kgalagadi South pastoral households cited shortage of drinking water for livestock in addition to stock theft, limited grazing areas and limited access to distant markets.

Drought

Droughts are common phenomena in semi-arid environments such as the Kalahari desert ecosystem of Botswana (Sandford 1979; Holm and Morgan 1985; Fako and Molamu 1995; Mogotsi et al. 2011a). Four broad categories of drought are recognized. Meteorological drought is usually measured by significant departures from normal precipitation observed over some period of time. Agricultural drought occurs when there is limited soil moisture to support crops and natural vegetation and results in less forage production than expected. Hydrological drought refers to deficiencies in surface and sub-surface water supplies to support established uses. Socio-economic drought is recognized when physical water shortage tangibly affects people in terms of their behaviour and options (Wilhite and Glantz 1985; IFAS 1998; Thurow and Taylor 1999). Whichever perspective one chooses to look at drought, the underlying cause is inadequate amount of moisture around which systems have stabilized.

Pastoral communities in rural Botswana have historically been and continue to be exposed to drought hazards (Holm and Morgan 1985; Mogotsi et al. 2011a). Expectedly, the two study areas of Ghanzi and Kgalagadi South both cited drought occurrence, irrespective of frequency and intensity, as a challenge. While other livelihood strategies like crop farming in higher rainfall areas view drought as a once-off temporary disruption, extensive pastoral systems are themselves inherently adapted to inevitable drought risks. This is achieved, in part, through risk-spreading and flexibility (Omosa 2005) such as livestock herd diversity and accumulation of large stock numbers so as to ensure long-term survival following drought losses (Barrow et al. 2007). In Botswana, past studies have estimated the minimum threshold number of animals in a herd to be no less than 40 (Carl Bro International 1982; CAR 2006) to start a self-sustaining subsistence cattle post. The complex role drought plays in such subsistence-based livelihoods in Botswana has been comprehensively elaborated (Mogotsi et al. 2013). A typical pattern usually emerges - inadequate rainfall amounts result in low natural pasture production on which livestock depend. Thus, areas in the proximity of livestock water points are overutilized and rapidly deteriorate, forcing animals to move further and further away in search of better grazing areas. The increased distance between better grazing areas and drinking water points ultimately takes its toll on livestock through loss of weight, decreased calving rates and, in extended drought periods, death of animals. The increased exposure to frequent drought episodes has the potential to erode resilience of pastoral systems and render them vulnerable (FAO 2009; Mogotsi et al. 2012).

Livestock Predation

The communal grazing area around Ghanzi borders other contrasting land use systems, mainly private freehold and leased ranches, WMAs and the Central Kalahari Game Reserve. With such close proximity to each other, the area experiences human-wildlife conflict. There is inevitable loss of livestock to predation by carnivores such as lions (Panthera leo), wild dogs (Lycaon pictus), brown hyenas (Parahyaena brunnea), spotted hyenas (Crocuta crocuta) and cheetahs (Acinonyx jubatus). The livestock-carnivore conflict is undoubtedly an emotive issue. Tourism generates significant revenue for Botswana's economy and is second only to the mining sector. Furthermore, wildlife tourism is seen as a viable diversification option away from the predominantly diamond mining-dependent economy. On the other hand, livestock are the backbone of subsistence pastoral livelihoods, and thus, losses, particularly from carnivore predation, are rarely tolerated or accepted (GoB 2005). To a small-scale farmer, loss of even one animal can translate into significant economic and social costs, while outsiders may view this simply on numerical terms and classify it as insignificant. All livestock are affected by predation, though calves, goats and sheep are particularly vulnerable and thus targeted. Also, it emerged that pastoral households whose animals are not protected at night by enclosures/ kraals or those with kraals that are not predator-proof suffer the most losses. If kraals are not properly constructed or maintained, predators can find an opening in the perimeter lining, or dig under the kraal and gain access, or simply climb/leap over. Hyenas were the predators singled out for causing the most livestock losses, and households showed greater animosity towards them. On some occasions, animals are targeted during daylight if no herding is done due to lack of reliable labour or if no guard dogs are used.

As part of a strategy to reduce and manage the conflict between humans and predators, the government through the Department of Wildlife and National Parks provides financial compensation to farmers who lose livestock to specified predators (GoB 2009b), even though amounts paid are generally deemed unsatisfactory (Kgathi et al. 2012). This discontent should be rectified accordingly since it is the individual pastoralists who are asked to sacrifice and compromise their livelihoods for the greater good of the nation and those who benefit directly from tourism. The adoption of such concepts as CBNRM acknowledges this point and strives to meet the needs of rural communities through sustainable economic exploitation of wildlife while achieving the country's resource conservation objectives. Niamir-Fuller et al. (2012) further assert that integration of wildlife management and livestock can offer multiple and complementary income sources, provided that stronger governance and regulatory co-management are in place.

Inadequate Drinking Water for Livestock

Pastoral livelihoods are climate dependent, driven particularly by rainfall in the semi-arid regions of Africa, the Near East, the New World and Australia (FAO 2001). In Botswana, the mean annual rainfall total ranges from less than 250 mm with a coefficient of variability of 45% in the extreme south-west part of the country to over 600 mm with a coefficient of variability of 25% in the extreme northeast (Bhalotra 1987). The low, erratic rainfall and associated excessive evaporation rates make water a limiting factor for pastoral communities of Ghanzi and Kgalagadi South. The area is subsequently devoid of large surface water bodies, save for the temporary water following summer rains in numerous pans dotted across the landscape. It is this patchiness of resources that has necessitated the opportunistic management that has kept the fragile and dynamic Kalahari ecosystem delicately balanced. Though underground water is accessible since the advent of borehole technology, there is still the added challenge of poor-quality water (high salinity) in some places (Hitchcock 1978; Mogotsi et al. 2011a). There is also the associated risk of hitting blanks, as well as debates on whether there is enough recharge to compensate for increased extraction of water in the Kalahari (Kgathi 1999; deVries et al. 2000; duPlessis and Rowntree 2003). The often prohibitive costs of drilling are beyond most individual pastoral households (CAR 2005), unless subsidized by the government (GoB 2002). Thus, as also revealed in this study, syndicates or farmer groups jointly own and manage boreholes or an individual borehole owner (with exclusive water rights) permits other users access to water. According to Nori et al. (2005), accessing resources and services of neighbours is a vital element for pastoralists, and therefore, the resource tenure system needs to be flexible enough to accommodate negotiations and arrangements among different groups and levels, depending on needs and on resources availability. Reciprocity is a valuable principle in such circumstances and represents some form of social capital.

Limited Grazing Land

Some Kgalagadi South pastoralists decried the inadequate grazing land for their livestock. Arid and semi-arid environments are climatically unpredictable, characterized by low net primary productivity and high resource variability. To counter this uncertainty, some form of pastoral mobility is essential to ensure that livestock have continuous access to adequate quality feed supplies and thus sustain the production system (Thomas et al. 2000). However, based on earlier assumptions that extensive communal pastoral systems were inefficient and irrational (Hardin1968), vast tracts of supposedly unused/empty land were expropriated and sedentarization was encouraged through modern fenced ranching systems. On the contrary, the unused/empty spaces were already inhabited by transhumant rural communities (Hitchcock 1978). As earlier noted, under the Tribal Grazing Land Policy of 1975, part of the communal land was demarcated into privately-owned ranches under 50-year leases, in a bid to combat perceived range degradation. The failures of TGLP are well documented (White 1993; Frimpong 1995) as sedentarization conflicts with migration as a strategy for coping with environmental variability and stress (Twyman 2001), including droughts.

Other competing economic activities like wildlife conservation and increasing number of human settlements in close proximity to communal grazing areas also require land. These activities, together with the weakening of traditional rangeland resource management systems, have stretched pastoral livelihoods in arid regions of Kgalagadi South, as they no longer have control over the shrinking land resource. Omosa (2005) argues that adequate resource access determines the sustainability of pastoralism, and by the same extension, it can be argued that pastoral households would have to somehow adapt or face destitution.

Loss of Livestock to Road Accidents

Ghanzi pastoral households experienced direct loss of livestock due to animal deaths or serious injuries caused by collisions with automobiles, especially along the recently constructed Trans Kalahari Highway cutting across southern Botswana and linking up the country with neighbouring South Africa to the south and Namibia in the west. While other main tarred roads (for example, the eastern A1 highway connecting the cities of Gaborone and Francistown) across the country are fenced to enhance road safety, the Trans Kalahari Highway is not, for fear of wildlife habitat fragmentation and disruption of migration and movement corridors. This situation in turn has contributed to an increased number of accidents involving vehicle collisions with domestic animals from the nearby cattle posts. According to ADBG (2011), data from the Ministry of Works, Transport and Communication reveal that there has been an increase in road accidents of approximately 64% along this Highway. The reasons for the increased road accidents are attributed to negligence, animals and mechanical deficiencies (ADBG 2011). During extended dry seasons and droughts, some animals may be drawn to the remnants of vegetation along the road reserves or lush green grass at the onset of the rainy season while some, especially donkeys and cattle, may sleep on the tarmac road presumably for warmth during winter nights. Furthermore, because of policy changes (for instance, Motor Vehicle Accident Fund and Road Traffic Act), owners may hesitate to claim livestock involved in road accidents as they are likely to be penalized for letting animals wander about and thus causing human injuries/deaths and/or damage to property. It is ironic that while physical infrastructure, particularly road networks, are meant to support other economic sectors including agriculture and tourism, it is the very same roads that are fragmenting habitats and literally killing domestic and wild animals.

Stray Animals

Stray animals were also a challenge to pastoral households in Ghanzi. This was attributed mainly to negligence on the part of herders, who ignore their responsibilities especially after receiving their monthly wages, then 'disappear' and only return after depleting their cash reserves. Absentee owners, often referred to as 'remote control' farmers, were also blamed for animals going astray as they spend most of their time in villages/towns under formal off-farm employment and rarely check on their livestock or the welfare of the herders. In the past, family labour was utilized in pastoral systems - a scenario which has changed as school-going children are no longer available to help out, as well as competition from other emerging sectors like construction for the limited labour pool. Moreover, the latter is now even more protected by the country's labour laws, something perceived in some quarters as a foreign concept which has prompted reluctant adjustments from some livestock owners.

Labourers were traditionally treated as part of the family; even though not adequately remunerated in cash, they were also paid in kind through free accommodation, food and other basic necessities enjoyed by the rest of the household members. In some cases, in return for their labour, herders could be given a few animals to progressively build their own herds. Already, as observed in the current study areas, there is general apathy towards menial jobs like livestock herding, especially among Batswana (citizenry of Botswana) youth. The continued availability of reliable labour is even more critical as pastoral households consist mostly of elderly people, who, even if willing, do no longer have the energy to personally look after their livestock. The ultimate fate of stray animals is governed by the Matimela Act (1969), which provides for the collection of stray livestock in Botswana's districts and the subsequent disposal of animals if unclaimed by the rightful owners after a stipulated period.

Livestock Theft

A related challenge experienced in both study areas is livestock theft, which was, however, more pronounced in Kgalagadi South rangelands. Livestock theft has been on the increase in Botswana (GoB 2011), mainly involving cattle, goats and donkeys. This crime is more prevalent in areas situated along the borders, like Kgalagadi South in proximity to South Africa and the periphery of cities and towns as in the case of Ghanzi. The latter area provides the market demand for livestock products, as is usual in most settlements with higher human populations. In the case of Kgalagadi South, it is mostly the dilapidated sections of fences along the border with South Africa that were pointed out as facilitating cross-border stock theft. In Botswana, theft of livestock can occur in several ways (GoB 2011), which include (1) killing lactating cows and stealing pre-weaned and unbranded calves, which are raised then later unlawfully branded and earmarked with the new owners' particulars, (2) unlawful branding of stray livestock which have joined part of the resident herd, (3) slaughtering for immediate consumption, usually individual offenders and (4) slaughtering stock usually by organized criminal syndicates to sell to unsuspecting or conniving butchery owners and bypassing the legal process of inspection of animals before slaughter at Council abattoirs as envisaged in the Livestock and Meat Industries Act (1962).

Because of the socio-economic importance of livestock in the country, the sector is arguably one of the most protected (Jefferis 2005) as all international trade in beef and cattle is prohibited as outlined in the Control of Livestock Industry Act (1968), except when routed through the Botswana Meat Commission (BMC). Traditionally, Batswana pastoral households used a wealth of knowledge acquired and refined over time to identify their animals by the use of marking techniques such as hot iron branding or ear marks and skin colour. While these are useful in identifying the owner of the animal(s), the system cannot adequately identify individual animals and hence the introduction of other techniques like ear tag numbers and lately the use of digital reticular bolus identification technology implemented under the Livestock Identification and Trace-back System. The latter is mostly compliance to regulatory requirements by Botswana's 'lucrative' European Union (EU) beef export market, which emphasizes stringent international standards for food quality and safety as well as disease control. The use of the bolus for curbing livestock theft was therefore an offshoot or by-product of the EU beef traceability requirement and is restricted only to cattle. The reticular bolus, however, was to be replaced with an electronic ear tag coupled with an analogue ear tag as an identification device effective the year 2013. Some farmers ingeniously use plastic ear tags to write down their contact details, usually mobile phone numbers, so that any of their lost animals which have been rounded up may be traced back to them through a simple phone call. All the preceding measures coupled with enforcement of the Stock Theft Act (1996) and other pertinent legislation have greatly eliminated commercialization of cattle raiding/rustling in Botswana, in contrast to other regions of Africa (Hendrickson et al. 1996; Kaimba et al. 2011), where the practice is widespread, sophisticated, more violent and destructive (Mkutu 2010; Schilling et al.2012).

Unfenced Grazing Areas

Unfenced grazing areas were cited as a challenge. At first, one can erroneously conclude that fencing of the commons is what pastoral communities are pining for, despite advice to the contrary (Behnke et al. 1993; Scoones 1995). But on closer scrutiny, the preceding challenges of livestock predation, stray animals and deaths of animals following collisions on highways and so on have part of their root causes in encroachment of pastoral rangelands by other, often competing, land users. The increased pressure on pastoral resources from other stakeholders can be alleviated, in the pastoralists' view, not only by virtual designation or zoning of land but also by physical barriers (i.e. fences) to secure their grazing areas from other users. This is perhaps a desperate response to the continued increase in external pressures on pastoral communities and thus the associated fear of their livelihoods being compromised. This is in line with observations made by Hartter et al. (2011), who opined that the presence or absence of resources influences behaviour and management as land use strategies are altered or maintained based on perceptions of risk and resource availability. The land tenure insecurity thus prevails among some pastoral households. As rightly observed by Pica-Ciamarra et al. (2010), 'it is not the system of land tenure that matters so much as the adequacy and security of tenure that allow for efficient and sustainable use of resources'. While there is provision for fencing of community grazing areas under the Agricultural Policy of 1991, implementation of the same is painstakingly slow.

Restricted Access to Markets

In an increasingly cash-driven global economy, contemporary pastoralists are forced to adapt or risk vulnerability and marginalisation (Nori et al. 2005). The growing demand for livestock products like meat and milk due to population growth, urban migration and urban income growth (Delgado et al. 1999) offers opportunities for pastoral communities in the so-called 'Livestock Revolution'. This is further supported by some governments, who often cite low livestock off-take rates and low average carcass weights in communal rangelands as evidence for room for improvement. And when markets continue to be 'underutilized' by pastoralists, then frustration sets in. The reasons for this 'underutilization' are well articulated by DFID (2011), among which are the poor fit between pastoralists' objectives in selling their livestock and the demands of the market, as well as long and risky marketing chains. Under traditional subsistence pastoral systems, livestock are commonly kept not to make money but rather to save money (Davies and Hatfield 2007). Furthermore, it is difficult to facilitate poverty alleviation and eradication when the pastoral communities themselves are economically isolated from large urban markets due to logistical and transport costs (Chadwick et al. 2008), as experienced by some households in Kgalagadi South. The BMC, which has the monopoly of all beef exports from Botswana, is located in Lobatse some 500 km from Tsabong in Kalagadi South, and thus, local butcheries are sometimes preferred even though they pay less or have little capacity to absorb high livestock sales. Kgosikoma (2006) found a similar trend in the same study area especially during drought years. In a study in East Africa by McPeak (2001), higher market access by pastoralists was associated with higher sales rates, while some used markets for destocking during drought periods and restocking after droughts.

While also far (approximately 640 km) from BMC headquarters, the Ghanzi area did not experience as much hardships in accessing markets due to BMC's Direct Cattle Purchase Scheme which buys cattle directly from farmers and therefore reduces transport costs. Another added advantage is the Trans Kalahari Highway. According to ADBG (2011), the profitability of transporting cattle from Ghanzi to Lobatse has increased significantly due to a reduction in cattle weight loss from 12% to 6% and bruising, as well as reduced travelling time. ADBG (2011) further estimated losses from the trekking of cattle from Ghanzi to Lobatse to have amounted to BWP 4.8 million to 7.2 million[b] (US$0.7 million to 1.0 million) per year before construction of the tarred road.

[b]US$1=6.97BWP.

That notwithstanding, some farmers still decry the high costs of transporting animals to the BMC abattoir in Lobatse. And because, for various reasons, the BMC can close and shut down operations for certain periods during the year, farmers are stuck with livestock even if they want to sell. This scenario in turn also lowers prices of animals due to limited selling outlets and thus translates into less profit for farmers. Other players ('speculators') also often come into the picture and buy young cattle cheaply, fatten them at feedlots and later sell them to the BMC and make higher returns. This usually affects and disadvantages pastoral households with smaller herds who have less bargaining power.

Sometimes the BMC abattoir can only absorb a fraction of cattle at a time, operating way below its full capacity. This quota limits the number of animals farmers can offload, and thus, it becomes even more costly to transport very few animals to Lobatse instead of a truck full of cattle to capitalize on economies of scale and maximize returns. Further compounding this problem for transport owners is the limited amount/absence of goods to carry back to pastoral areas after offloading cattle at the abattoir instead of trucks returning empty. This situation in turn hurts the BMC, as it becomes less efficient and in the process incurs high overhead costs per unit of throughput due to low supply of slaughter animals.

Improved access to markets also has other added benefits for pastoral households apart from livestock sales, such as ease of importing commercial livestock feed during extended periods of drought, securing livestock drugs and vaccines as well as borehole equipment and spares.

Complex System, No Easy Fix

In light of the preceding constraints, finding clear-cut mitigation measures is a daunting task in a complex ecological and socio-economic system. Nonetheless, an attempt at some policy considerations is made.

Communal Pastoral System is here to Stay

The communal pastoral system in Botswana needs to be acknowledged as a self-sustaining livelihood strategy in supposedly marginal land, complete with its own ecological, biological and social dynamics. Once this is recognized, then policies which continuously strive to 'rescue' and 'modernize' pastoral communities will be reconsidered. For example, current policy will push for some form of exclusion (fencing) when confronted with the preceding challenges in the study areas such as stray animals, livestock predation, stock theft, involvement of livestock in road accidents, livestock diseases and uncontrolled breeding. This strategy, however, will play right into the hands of proponents of fencing of the commons. Instead, policy should 'think beyond the fence' and lobby for the resuscitation of traditional natural resource management institutions at the local level, with more jurisdictions over land use. Also, to avoid the continued underestimation of the importance of livestock in national statistics, the capture of field data should also be modified to include non-monetary uses of livestock at the local level, such as animals used in paying bride price (bogadi) or goats slaughtered for home consumption or social events.

Improve Land Tenure Security

Further complicating the current policy drive is that Batswana rural communities (and still some urbanites) are entrenched in livestock farming, so there simply is not enough land to be demarcated into viable ranches for everyone, not now and certainly not in future. Communal pastoral land is further shrinking through demarcation for private livestock and wildlife ranching, crop farming, human settlements and other uses. As it is now, the push for fencing of the commons is so strong that even some communal pastoral households aspire to own or lease private ranches, if not for production then certainly for the prestige attached to it. This in itself is a reflection of the insecurity associated with communal land tenure.

Increase Involvement of Stakeholders

While commendable effort has been made to provide pastoral fora, mainly through groupings of livestock producers or Associations in Botswana, there is need to continuously involve the poorer pastoralists and avoid hijacking of such associations by richer and more powerful pastoralists. That way, a more representative voice of pastoral challenges and aspirations can be heard. Worryingly, the non-governmental organizations (NGOs) in Botswana rarely advocate for preservation of communal pastoral land use. Instead, topical issues for NGOs are wildlife-related and occasionally livestock health, with the latter often emphasizing the need for proper 'modern' management practices. This should be rectified. Financial challenges for pastoralists like expensive borehole drilling and livestock feeds can be alleviated through increased access to credit from financial institutions, which also need to come on board and tailor their products to suit the complexity of pastoral systems.

Restructure and Facilitate Access to Markets

While not discouraging high-value beef export market to Europe, the Policy Framework for Pastoralism in Africa (AU 2010) stresses the need to pay more attention to markets within Africa and to foster regional trade, as well as tapping into the growing markets in the Middle East and Southeast Asia. These markets are relatively less stringent in terms of animal husbandry standards which make it less expensive and therefore easier for small-scale pastoralists to sell. And the timing is apt. For example, local beef consumption in Botswana has risen over the years and gradually reduced BMC's market share from 80% in 1981 to 44% in 2002 (CAR 2005). Considering that despite (or because of) government protection, BMC is running aground after years of posting record losses, a thorough overhaul and restructuring of the BMC is needed, or alternatively BMC's monopoly should be ended and the beef industry market liberalized - whichever is in the best interests of the livestock industry. Another market-related point is that pastoral households, especially those with fewer herds, tend to sell in order to have immediate cash in hand and do not have the luxury of applying for quotas to sell animals beforehand or waiting for days for the BMC to process their payments after selling (often cheques or electronic money transfers into the sellers' bank accounts).

The Policy Framework (AU 2010) further emphasizes the need to retain value locally through more processing of livestock products where economically viable - opportunities which pastoral communities can benefit from while the government can increase its revenue.

Value Indigenous Knowledge

Resilient pastoral communities possess a body of knowledge which has evolved over the years. Lately, research institutions in Botswana (e.g. The Centre for Scientific Research, Indigenous Knowledge and Innovation hosted by the University of Botswana) and the government (primarily through the Ministry of Infrastructure, Science and Technology) have demonstrated keen interest in unearthing and capturing indigenous technical knowledge of rural communities, as well as developing a policy on indigenous knowledge systems. This is a positive development, though still in its infancy. The accumulated and refined knowledge base includes, but is by no means restricted to, weather forecasts and precipitation indicators, livestock health problems, important poisonous or injurious plants, range degradation and drought mitigation strategies (e.g. Dube and Sekhwela2008;Cassidy et al. 2011; Mogotsi et al. 2011b, Motlhanka and Nthoiwa 2013). Pastoral communities also traditionally had safety nets for the less privileged members of their society. If policy can tap into these indigenous knowledge systems, support and strengthen them, then more effective and efficient use of natural resources can be realized.

Develop Area-Specific Disaster Management Systems

Pastoral areas are not immune to uncontrolled veld fires, droughts, floods and epidemic diseases. In the case of Kgalagadi and Ghanzi where recurrent droughts are common, interventions ought to be tailored to local conditions. The current 'blanket' aid programmes which might be appropriate in one Sub-district may not necessarily reduce household vulnerability to drought in another area and might instead reinforce the negative effects of drought (Mogotsi et al. 2012). For example, livestock feed subsidies may not be the most appropriate response in an area experiencing serious shortage of drinking water for livestock. Nor is subsidizing prices of concentrates desirable, when there is not enough roughage to sustain animals through the drought. Nor is encouraging destocking effective through increase of BMC prices, when already-weak animals have to be moved long distances and quarantined in some cases before reaching the market. Also, death of some animals during droughts should not always be viewed with alarm, as this is a natural regulatory mechanism of the semi-arid pastoral system. Interestingly, it may also be economically sensible for the farmer to let the animal die rather than to try to save it. But by making available highly subsidized commercial livestock feed during droughts, the government is inadvertently encouraging pastoralists to keep an artificial high number on a stressed ecosystem. Instead, if traditional community-specific and modern disaster management approaches are merged, then better early-warning, actual-event and post-disaster response measures can be formulated.

CONCLUSIONS

Extensive pastoral systems of the Kalahari remain central to rural livelihoods. The challenges presented in this study act concomitantly to exert pressure on, and erode resilience of, pastoral households and subsequently expose the system to risk. If these issues are honestly recognized and addressed through deliberately targeted policy interventions, then the long-term sustainability of extensive pastoralism can be assured, failing which, the non-pastoral sectors should be ready and willing to find an alternative, albeit more expensive, economic activity to support the 'once upon a time' pastoral population. Prospects for success of extensive pastoral systems in semi-arid regions such as Botswana lie in the overriding importance of collective management and flexibility, more so in a changing global climate.

AUTHORS' CONTRIBUTIONS

DM conceived the study idea, collected and interpreted data as well as drafting of the manuscript. KM improved the initial draft, interpreted data and drafted the manuscript. Both authors read and approved the final manuscript.

REFERENCES

1. Abel, N. 1997. Mis-measurement of the productivity and sustainability of African communal rangelands: a case study and some principles from Botswana. Ecological Economics 23: 113–133.

2. ADBG. 2011. Botswana: Trans-Kgalagadi road project. Project performance evaluation report.Operations Evaluation Department, Project and Program Evaluation Division, African Development Bank Group. http://www.afdb.org/fileadmin/uploads/afdb/Documents/Evaluation Reports/Botswana-PPER_FINAL-May%20Final.pdf Accessed 31 Jan 2013
3. APRU. 1980. Beef production and range management in Botswana. Gaborone: Animal Production and Range Research Unit, Ministry of Agriculture.
4. AU. 2010. Policy framework for pastoralism in Africa: Securing, protecting and improving the lives, livelihoods and rights of pastoralist communities. Addis Ababa: Department of Rural Economy and Agriculture, African Union.
5. Barrow, E, J Davies, S Berhe, V Matiru, N Mohamed, W Olenasha, and M Rugadya. 2007. Pastoralists as shrewd managers of risk and resilience in the Horn of Africa. Policy Brief No. 1 of 5. Nairobi: IUCN Eastern Africa Regional Office.
6. Behnke, R. 1985. Measuring the benefits of subsistence versus commercial livestock production in Africa. Agricultural Systems 16 (2): 109–135.
7. Behnke, R. 2008. The economic contribution of pastoralism: Case studies from the Horn of Africa and southern Africa. Nomadic Peoples 12: 45–79.
8. Behnke, RH, I Scoones, and C Kerven. 1993. Range ecology at disequilibrium: New models of natural variability and pastoral adaptation in African Savannas. London: Overseas Development Institute.
9. Bhalotra, YPR. 1987. The drought of 1981–87 in Botswana. Gaborone: Department of Meteorological Services.
10. BIDPA. 2006. Prospects for export diversification in Botswana. Botswana Institute for Development Policy Analysis/World Bank report. http://www.bidpa.bw/img_upload/pubdoc_20.pdf Accessed 10 Apr 2013
11. Burgess, J. 2002. Botswana: Country pasture/forage resource profile. Food and Agriculture (FAO) report. http://www.fao.org/ag/AGP/AGPC/doc/Counprof/Botswana/botswana.htm Accessed 19 Apr 2012

12. CAR. 2005. Crop-livestock systems development in southern Africa: Priorities for future research. Botswana country report. Gaborone: Centre for Applied Research.
13. CAR. 2006. Environmental assessment of Botswana's livestock sector. Centre for Applied Research report. Ministry of Finance and Development Planning: Gaborone.
14. Carl Bro International. 1982. An evaluation of livestock management and production in Botswana with special reference to the communal areas, vol. 1–111. Gaborone: Evaluation Unit, Ministry of Agriculture and the Commission of the European Communities European Development Fund.
15. Cassidy, L, J Wilk, DL Kgathi, H Bendsen, BN Ngwenya, and K Mosepele. 2011. Indigenous knowledge, livelihoods and government policy in the Okavango Delta, Botswana. In Rural livelihoods and political economy, ed. Kgathi DL, Ngwenya BN, Wilk J, 75–97. New York: Nova Science Publishers.
16. Chadwick, K, J Otte, and D Roland-Holst. 2008. Information failures in livestock markets: Evidence from Lao PDR. Pro-Poor Livestock Policy Initiative.A Living from Livestock.Research report. Rome: FAO.
17. Coppock, DL, and S Sovani. 1999. Is supplementation justified to compensate pastoral calves for milk restriction? Journal of Range Management 52: 208–217.
18. Darkoh, MBK, and JE Mbaiwa. 2003. Globalization and the livestock industry in Botswana. Singapore Journal of Tropical Geography 23: 149–166.
19. Davies, J, and R Hatfield. 2007. The economics of mobile pastoralism: A global summary. Nomadic Peoples 11: 91–116.
20. Delgado, C, M Rosegrant, H Steinfeld, S Ehui, and C Courbois. 1999. Livestock to 2020: The next food revolution. 2020 vision discussion paper. Washington, DC: International Food Policy Research Institute.
21. deRidder, N, and KT Wagenaar. 1986. Energy and protein balances in traditional livestock systems and ranching in eastern Botswana. Agricultural Systems 20: 1–16.
22. deVries, JJ, ET Selaolo, and HE Beekman. 2000. Groundwater recharge in the Kalahari, with reference to paleo-hydrologic conditions. Journal of Hydrology 238: 110–123.

23. DFID. 2011. Pastoralism: Progressing policies that favour pastoralists. Facilitating access to world markets.Department for International Development. http://www.dfid.gov.uk/r4d/Output/187272/Default.aspx Accessed 18 Dec 2012
24. Dube, OP, and MBM Sekhwela. 2008. Indigenous knowledge, institutions and practices for coping with variable climate in the Limpopo Basin of Botswana. In Climate change and adaptation, ed. Leary N, Adejuwon J, Barros V, Burton I, Kulkarni J, Lasco R, 71–89. London: Earthscan.
25. duPlessis, AJE, and KM Rowntree. 2003. Water resources in Botswana with particular reference to the Savanna regions. South African Geographical Journal 85: 42–49.
26. Elhadi, YAM, DM Nyariki, VO Wasonga, and WN Ekaya. 2012. Transient poverty among pastoral households in the semi-arid lowland of Baringo district, Kenya. Ozean Journal of Social Sciences 5: 9–19.
27. ESAP. 2003. Challenges and opportunities of livestock marketing in Ethiopia. In Proceedings of the 10th annual conference of the Ethiopian Society of Animal Production (ESAP) held August 22–24 in Addis Ababa, Ethiopia, ed. Jobre Y, Gebru G,
28. Fako, TT, and L Molamu. 1995. The seven-year drought, household food security and vulnerable groups in Botswana. Pula: Botswana Journal of African Studies 9: 48–70.
29. FAO. 2001. Pastoralism in the new millennium. FAO Animal Production and Health Paper 150. http://www.fao.org/docrep/005/y2647e/y2647e00.htm#toc Accessed 9 Jan 2013
30. FAO. 2009. Review of evidence on drylands pastoral systems and climate change: Implications and opportunities for mitigation and adaptation. Land and Water Discussion Paper 8. Rome: FAO.
31. Flintan, F, and A Cullis. 2010. Introductory guidelines to participatory rangeland management in pastoral areas. Addis Ababa: Save the Children/US.
32. Fraser, EDG, AJ Dougill, WE Mabee, M Reed, and P McAlpine. 2006. Bottom up and top down: analysis of participatory processes for sustainability indicator identification as a pathway to community empowerment and sustainable environmental management. Journal of Environmental Management 78: 114–127.

33. Frimpong, K. 1995. A review of the tribal grazing land policy in Botswana. Pula: Botswana Journal of African Studies 9: 1–16.

34. GoB. 1975. National policy on tribal grazing land. Paper No. 2 of 1975. Gaborone: Government of Botswana.

35. GoB. 1991. National policy on agricultural development. Gaborone: Ministry of Agriculture, Government of Botswana.

36. GoB. 1999. Community-based natural resource management practitioners guide. Gaborone: Department of Wildlife and National Parks, Government of Botswana.

37. GoB. 2002. Revised national policy for rural development. Ministry of Finance and Development Planning, White Paper No. 3. Gaborone: Government of Botswana.

38. GoB. 2005. Wildlife Conservation and National Parks (Lions) (Killing Restriction) Order. Gaborone: Department of Wildlife and National Parks, Government of Botswana.

39. GoB. 2009a. Final report on establishment of the Botswana vulnerability assessment and analysis system for national food security. Gaborone: Government of Botswana.

40. GoB. 2009. New compensation policy. Gaborone: Department of Wildlife and National Parks, Government of Botswana.

41. GoB. 2011. Crime prevention guidelines for livestock. Gaborone: Government of Botswana. http://www.gov.bw/en/Ministries--Authorities/Ministries/State-President/Botswana-Police-Service-/About-the-BPS/Stock-Theft-Branch/ Accessed 12 Jan 2013

42. GoB. 2013. Annual agricultural surveys preliminary results for 2011. Gaborone: Government of Botswana. http://www.cso.gov.bw Accessed 7 June 2013

43. Greenhow, T. 1978. The Tribal Grazing Land Policy and integrated land-use planning: A district view. Botswana Notes and Records 10: 159–168.

44. Hardin, G. 1968. The tragedy of the commons. Science 162: 1243–1248.

45. Hartter, J, A Goldman, and J Southworth. 2011. Responses by households to resource scarcity and human-wildlife conflict: Issues of fortress conservation and the surrounding agricultural landscape. Journal for Nature Conservation 19: 79–86.

46. Hendrickson, D, J Armon, and R Mearns. 1996. Livestock raiding among the pastoral Turkana of Kenya: Redistribution, predation and the links to famine. Institute of Development Studies Bulletin 27: 17–30.
47. Hitchcock, RK. 1978. Kalahari cattle posts: A regional study of hunter-gatherers, pastoralists and agriculturalists in the Western Sandveld Region, vol. I. Preliminary draft. Gaborone: Ministry of Local Government and Lands.
48. Holm, JD, and RG Morgan. 1985. Coping with drought in Botswana: An African success. Journal of Modern African Studies 23: 463–482.
49. IFAS. 1998. The disaster handbook. National Edition Publication DH 1605. Gainesville: Institute of Food and Agricultural Sciences, University of Florida.
50. Jefferis, K. 2005. How trade liberalization can contribute to revitalise Botswana's beef and cattle sector. Proceedings of the Botswana Agricultural Research Information Network (BARIN) conference, 24–26 May 2005, Sebele, Botswana.
51. Kaimba, GK, BK Njehia, and AY Guliye. 2011. Effects of cattle rustling and household characteristics on migration decisions and herd size amongst pastoralists in Baringo District, Kenya. Pastoralism: Research, Policy and Practice 1: 18.
52. Keijsper, V. 1993. Livestock development in Botswana: Consequences for the rural poor. Africa Spectrum 28: 323–339.
53. Kgathi, DL. 1999. Water demand, population and sustainability in Botswana: Implications for development policy. A paper prepared for the Population, Development and Environment Project. Luxembourg: International Institute for Applied Systems Analysis.
54. Kgathi, DL, G Mmopelwa, B Mashabe, and K Mosepele. 2012. Livestock predation, household adaptation and compensation policy: A case study of Shorobe village in northern Botswana. Agrekon: Agricultural Economics Research, Policy and Practice in Southern Africa 51: 22–37.
55. Kgosikoma, OE. 2006. Effects of climate variability on livestock population dynamics and community drought management in Kgalagadi, Botswana. MSc thesis. Norwegian University of Life Sciences, Norway.

56. McPeak, J. 2001. Pastoralists' use of markets. Pastoral Risk Management Project Research Brief 01-04-PARIMA. Davis: Global Livestock Collaborative Research Support Program.
57. Mkutu, KA. 2010. Complexities of livestock raiding in Karamoja. Nomadic Peoples 14: 87–105.
58. Mogotsi, K, MM Nyangito, and DM Nyariki. 2011. The perfect drought? Constraints limiting Kalahari agro-pastoral communities from coping and adapting. African Journal of Environmental Science and Technology 5: 168–177.
59. Mogotsi, K, AB Moroka, O Sitang, and R Chibua. 2011. Seasonal precipitation forecasts: Agro-ecological knowledge among rural Kalahari communities. African Journal of Agricultural Research 6: 916–922.
60. Mogotsi, K, MM Nyangito, and DM Nyariki. 2012. Vulnerability of rural agro-pastoral households to drought in semi-arid Botswana. Livestock Research for Rural Development24(10),Article #183. http://www.lrrd.org/lrrd24/10/mogo24183.htm Accessed 15 December 2012
61. Mogotsi, K, MM Nyangito, and DM Nyariki. 2013. The role of drought among agro-pastoral communities in a semi-arid environment: The case of Botswana. Journal of Arid Environments 91: 38–44.
62. Motlhanka, DMT, and GP Nthoiwa. 2013. Ethnobotanical survey of medicinal plants of Tswapong North, in Eastern Botswana: A case of plants from Mosweu and Seolwane villages. European Journal of Medicinal Plants 3: 10–14.
63. Mrema, M, and S Rannobe. 1996. Goat production in Botswana: Factors affecting production and marketing among small-scale farmers. In Small ruminant research and development in Africa. Proceedings of the third biennial conference of the African Small Ruminant Research Network, Kampala, Uganda, ed. Lebbie SHB, Kagwini E, 105–109. Nairobi: International Livestock Research Institute (ILRI). http://www.fao.org/wairdocs/ilri/x5473b/x5473b0v.htm Accessed 17 November 2012
64. Niamir-Fuller, M, C Kerven, R Reid, and E Milner-Gulland. 2012. Co-existence of wildlife and pastoralism on extensive rangelands: Competition or compatibility? Pastoralism: Research. Policy and Practice 2: 8.

65. Nori, M, J Switzer, and A Crawford. 2005. Herding on the brink: Towards a global survey of pastoral communities and conflict.An occasional paper from the IUCN Commission on Environmental, Economic and Social Policy; Gland (Ch). http://www.iisd.org/pdf/2005/security_herding_on_brink.pdf Accessed 4 October 2012

66. Omosa, EK. 2005. The impact of water conflicts on pastoral livelihoods: The case of Wajir District in Kenya. Winnipeg: International Institute for Sustainable Development.

67. Pica-Ciamarra, U, J Otte, and C Martini. 2010. Livestock sector policies and programmes in developing countries - a menu for practitioners. Food and Agriculture Organization (FAO): Rome.

68. Sandford, S. 1979. Toward a definition of drought. In proceedings of the symposium on drought in Botswana. Gaborone: Botswana Society.

69. SAS. 2005. SAS/STAT user's guide. Cary: SAS (Statistical Analysis Systems) Institute Inc.PubMed Abstract

70. Schilling, J, FEO Opiyo, and J Scheffran. 2012. Raiding pastoral livelihoods: Motives and effects of violent conflict in north-western Kenya. Pastoralism: Research, Policy and Practice 2: 25.

71. Scoones, I. 1995. New directions in pastoral development in Africa. In Living with uncertainty: New directions in pastoral development in Africa, ed. Scoones I, 1–36. London: Intermediate Technology Publications Ltd..

72. Thomas, DSG, and D Sporton. 1997. Understanding the dynamics of social and environmental variability: The impacts of structural land use change on the environment and peoples of the Kalahari, Botswana. Applied Geography 17: 11–27.

73. Thomas, DSG, D Sporton, and J Perkins. 2000. The environmental impact of livestock ranches in the Kalahari, Botswana: Natural resource use, ecological change and human response in a dynamic dryland system. Land Degradation and Development 11: 327–341.

74. Thurow, TL, and CA Taylor Jr. 1999. Viewpoint: The role of drought in range management. Journal of Range Management 52: 413–419.

75. Tsimako, B. 1991. The Tribal Grazing Land Policy ranches: Performance to date. Gaborone: Ministry of Agriculture.
76. Twyman, C. 2001. Natural resource use and livelihoods in Botswana's Wildlife Management Areas. Applied Geography 21: 45–68.
77. White, R. 1993. Livestock development and pastoral production on communal rangeland in Botswana. Gaborone: The Botswana Society.
78. Wilhite, DA, and MH Glantz. 1985. Understanding the drought phenomenon: The role of definitions. Water International 10: 111–120.

Chapter 7

Natural Conservationists? Evaluating the Impact of Pastoralist Land Use Practices on Tanzania's Wildlife Economy

Fred Nelson

Maliasili Initiatives, P.O. Box 293, Underhill, VT, 05489, USA

ABSTRACT

The land management practices of pastoralist Maasai communities have a major bearing on landscapes and wildlife habitats in northern Tanzania and play a key role in maintaining habitat for one of the world's most spectacular assemblages of terrestrial large mammals. Pastoralists manage lands according to locally devised rules designed to manage and conserve key resources such as pastures and water sources. Dry season grazing reserves are an important part of traditional land management systems in many pastoralist communities, providing a 'grass bank' for livestock to consume during the long dry season when forage invariably becomes scarce and domestic animals are stressed for water and nutrients. Because of the scale and importance of northern Tanzania's wildlife-based tourism industry, and its indirect dependence on communal lands under the authority of pastoralists, these land use practices have an important economic dimension. By conserving large proportions of northern Tanzania's wildlife ecosystems, local pastoralist communities collectively make an important contribution to the national and regional economy. Using data regarding the degree to which wildlife depends on pastoralist lands in different ecosystems, and the relative importance of different areas in terms of generating revenue for the northern safari circuit, the annual value of pastoralist land uses to the wildlife-based tourism industry in northern Tanzania is estimated at approximately US $83.5 million. The economic value of pastoralists' contribution to wildlife conservation highlights the importance of Tanzanian policies in land, livestock, tourism, and wildlife sectors prioritizing measures that promote communal rangeland management and support traditional land use practices.

PASTORALISTS, WILDLIFE, AND TOURISM IN NORTHERN TANZANIA: AN OVERVIEW

Northern Tanzania's savannahs and grasslands contain some of the world's most diverse and abundant remaining terrestrial wildlife populations. This abundance of wildlife has, during the past 20 years, been the key asset underlying the development of one of Africa's fastest-growing tourism industries. By 2007, tourism was estimated to contribute $1.6 billion directly and indirectly to Tanzania's economy, accounting for about 11% of the country's Gross Domestic Product (Mitchell et al. 2009). Northern Tanzania's tourism circuit, stretching between Mount Kilimanjaro on the east and the Serengeti to the west, generates more than half of Tanzania's foreign earnings from tourism, making it 'one of the most valuable strips of tourism real estate in Africa' (Mitchell et al. 2009, page 7).

In order to conserve wildlife and their habitats, and to develop sites for tourism activities, a large number of state-protected areas - national parks and game reserves - have been established in northern Tanzania's savannah landscapes during the course of the last 50 years (and in a few cases dating back to earlier colonial-era measures). These state-protected areas provide the core attractions and infrastructure of the northern tourism circuit (Figure 1).

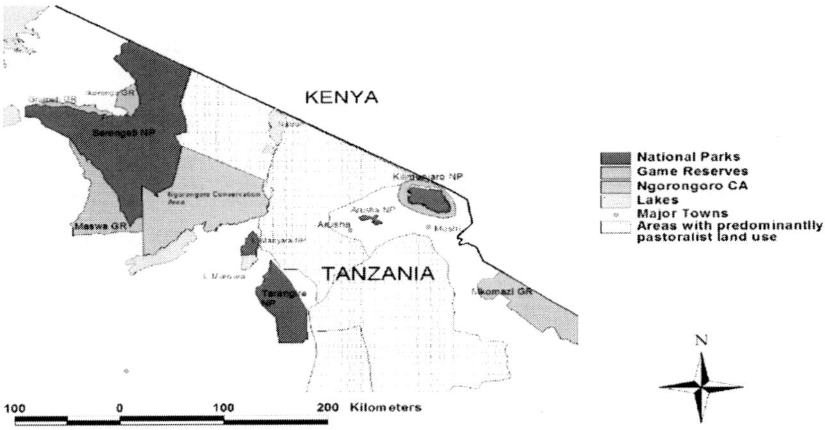

Figure 1: Main state-protected areas and tourism attractions of the northern Tanzanian safari circuit, and areas with predominantly pastoralist land uses.

The areas that are now designated as state-protected areas were formerly used and occupied by local communities, with parks such as Serengeti and Tarangire occupied by pastoralists prior to their gazettement. When the colonial and post-independence governments established national parks and game reserves in these customarily owned lands, local people were evicted and rights of occupancy gradually extinguished. Despite this long-running effort to divide the landscape into areas for people and areas for wildlife conservation, northern Tanzania continues to be characterized by a high level of interaction and co-existence between local resident people and wildlife. Parks and reserves are invariably situated within larger ecosystems, and wildlife continues to move across the artificial boundaries dividing communal and state lands. Many species of large mammals, such as elephants, wildebeest, and zebra, undertake annual migrations based on the seasonal availability of water, forage, and nutrients. Wildlife in northern Tanzanian ecosystems continues to rely on many lands and resources which are located outside protected areas on private and community lands, making these latter areas essential for the long-term conservation of wildlife.

The ways that local pastoralist communities manage their lands and resources thereby have a major impact on the conservation of wildlife in savannah ecosystems in northern Tanzania. Indeed, pastoralist land management practices such as grazing and the use of fire have had a major influence over the course of the past 3,000 to 4,000 years in shaping East African savannah landscapes. Many observers attribute the contemporary abundance of wildlife in East Africa to the historic influences of pastoralists on savannah landscapes and on the general ecological compatibility between pastoralists and wild large mammals (Collet 1987; Western 1989; Homewood and Rodgers 1991). Pastoralists in northern Tanzania's savannah ecosystems thereby provide an economically valuable ecological service by conserving wildlife on their lands, which in turn helps to sustain the natural assets that Tanzania's growing tourism industry depends on. This article describes the nature and extent of that ecological service by pastoralists and develops a general and approximate estimate of its economic value in the context of Tanzania's wildlife-based tourism industry. Such an understanding is central to the development of economically efficient and productive national policies in sectors such as land, livestock, and wildlife in Tanzania and the harmonization of rural livelihoods with national and international conservation objectives.

Sharing the Land: Pastoralist and Wildlife Interactions

It has long been recognized that pastoralists and their livestock, and wildlife live alongside one another in East African savannah ecosystems, where they exhibit a high degree of spatial overlap or co-existence (Sitters et al. 2009; du Toit 2011). Both pastoralist livestock and wildlife utilize foraging strategies based on mobility (seasonal migrations or transhumance) in order to access pasture and water occurring in an unpredictable physical environment (Homewood and Rodgers 1991).

The degree to which pastoralists' livestock and wild ungulates compete for grazing and water resources, which in turn shapes the potential for co-existence or even symbiotic interactions, has been the subject of both research and debate within East Africa for decades. In the 1950s and 1960s, the general perception was that pastoralist overgrazing was prevalent and resulted in environmental degradation and damage to wildlife habitat. This overgrazing narrative was a central factor in the creation of the Serengeti National Park and relocation of resident people, including about 1,000 Maasai pastoralists, in 1959 (Collet 1987; Neumann 1998; see also Grzimek and Grzimek 1960). A core concept of range science at this time was the concept of carrying capacity holding that a given unit of land area can only support so many livestock (or wild animals) due to the limitations of a given volume of forage which that land area can produce (Pratt and Gwynne 1977). Pastoralists, according to this paradigm, would exceed the carrying capacity of the land by producing more livestock over time, but not selling (or eating) sufficient numbers of animals, leading to more grazing pressure on limited forage resources. Underlying this process of overgrazing was the so-called cattle complex, meaning the cultural value pastoralists place on building up large herds of livestock, and the tragedy of the commons whereby communally owned and managed pastures lacked controls limiting individuals' livestock numbers and consumption of forage, making progressive overgrazing seemingly inevitable (Lamprey 1983; see alsoHardin 1968).

Many of these ideas about overgrazing, the cattle complex, and a tragedy of the commons on communal lands continue to inform policy debates over livestock, land tenure, and natural resource management in East Africa today (Mattee and Shem 2006). However, during the last 30 years, a wide range of ecological, anthropological, and institutional studies have altered thinking about rangelands and discredited or modified many earlier assumptions about pastoralist land management practices. Ecological research in semi-arid areas across Africa has demonstrated the key role that rainfall plays in the condition of above-ground vegetation; during droughts, areas tend to lose cover but recover fairly rapidly following rainfall (Behnke et al. 1993; Dodd 1994; Scoones 1995).

Pastoralists and the Environment: Managing Uncertainty

Pastoralists respond to these highly variable and unpredictable, non-equilibrium conditions by moving between different local pastures depending on where rain has fallen at a given time, and maintaining reciprocal relationships with neighbouring and distant communities so that livestock may seek refuge or access pastures in areas where conditions are better during times of drought (Ellis et al. 1988). A considerable body of research has emerged during the last 30 years demonstrating that pastoralist management strategies such as mobility, rotational pasture use through different wet season and dry season grazing areas, and building up cattle numbers so as to be able to endure catastrophic but relatively common periods of drought are ecologically rational measures for coping with the variable semi-arid and arid environments that these people live in (Homewood and Rodgers 1991; Behnke et al. 1993; Scoones 1995; Bourn and Blench 1999).

A major change has also occurred in thinking about the 'tragedy of the commons,' both in general and specifically in relation to pastoralists. Hardin's (1968) central assumption was that such commons are effectively ungoverned by the community of resource users which utilizes such areas. By 1990s, though an array of research (e.g. Berkes 1989; Ostrom 1990) was documenting that, in fact, many commons around the world are subject to locally devised and enforced rules that govern the use of shared resources. In many instances, local rules governing resource use are as or more effective at conserving resources such as forests than the state-protected areas that formal conservation efforts often promote (Hayes 2006; Ostrom and Nagendra 2006). East African pastoralists generally manage rangelands collectively through community-level institutions (Barrow 1988; McCabe 1990; Homewood and Rodgers 1991). Maasai communities traditionally divide their collective pastures into different types of areas governed by certain traditional rules. The most fundamental division is of pastures used during the wet season and areas used as dry season refuges. Wet season pastures tend to be those without permanent water that can only be accessed during the periods of rainfall. Dry season pastures tend to be areas which have permanent water nearby or which are particularly resistant to drought, such as forests which retain vegetation that livestock can access even during severe droughts. In addition, some pastures are designated for use by calves and weak or infirm livestock. These areas, called *olokeri*, are close to permanent settlements and sources of water and are allocated to particular households rather than being managed at the community or village level. Figure 2 illustrates this traditional land use scheme as mapped and formalized in village by-laws in one pastoralist village in Loliondo Division, northern Tanzania (UCRT 2010).

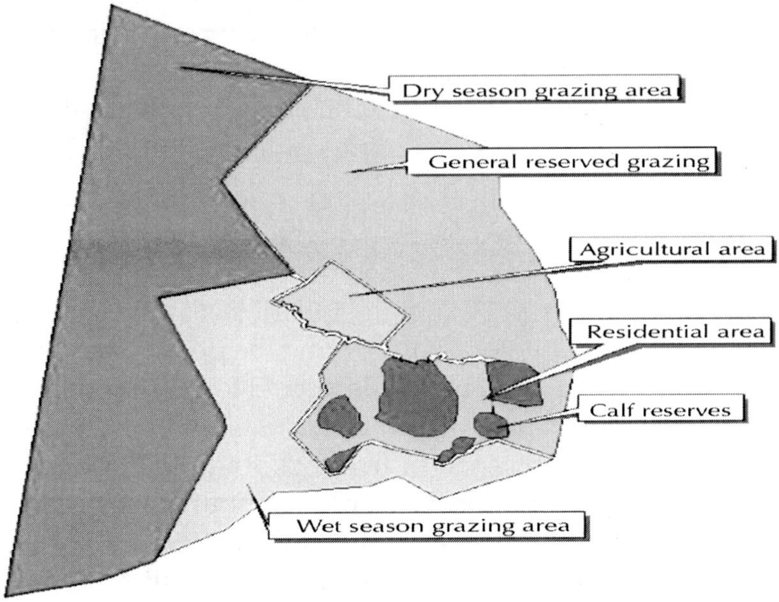

Figure 2: Land use zones for a pastoralist village in Loliondo, northern Tanzania**.** This is based on traditional pasture management system and formalized through land use plans and village by-laws. Adapted from UCRT (2010).

Pastoralist-Wildlife Interactions

In terms of better understanding livestock-wildlife interactions in East African savannahs, a starting point is the reality that different types of animals eat different types of forage (du Toit 2011). Species such as giraffe, gerenuk, kudu, and domestic goats are browsers, eating leaves from trees and bushes, whereas species such as wildebeest, zebra, buffalo, gazelle, and cattle are grazers. Bell (1971), based on research carried out in the Serengeti, describes a process of 'grazing succession' whereby different species of wild ungulates, such as wildebeest and zebra, progressively move through a grazing area. The diversity of grazing ungulates that exists in East Africa is possible because different species can co-exist without directly competing for grazing since their feeding preferences are different. For example, zebra eat taller, more coarse grass while wildebeest, as well as smaller grazers such as gazelle, are more selective and efficient grazers (antelopes and gazelle, unlike zebra, are ruminants and have much more efficient digestive systems) preferring shorter swards of grass. This enables wildebeest to graze areas that zebra have already passed through and grazed over initially, and in this way, zebra may facilitate grazing by wildebeest and other more selective grazers including cattle (du Toit 2011).

Interactions between livestock and wildlife may, as conditions change, be both competitive or, as in the process of grazing succession, facilitative depending on the species involved and the seasonal availability of resources. For example, Odadi et al. (2011) describe how zebra and cattle compete for forage during the dry season, when forage is scarcer, while zebra facilitate cattle access to grazing during the wet season, when grass growth is at its high point. It is important to highlight that it is not only the presence of livestock, but their absence as well, that shapes habitat conditions for wildlife in both positive and negative ways. As an example of the latter, Gichohi's (1992) study of Nairobi National Park in Kenya describes how after pastoralists were excluded from the park in the 1960s, the lower levels of cattle grazing as well as the use of fire, resulted in reduced numbers inside the park of selective grazers which prefer a low grass sward; this included hartebeest, warthog, and Thomson's gazelle. These species had benefited from the presence of livestock and pastoralist land management practices. This work illustrates what Western and Gichohi (1994) term as 'segregation effects' in savannah ecosystems, whereby human interventions that remove people from areas where they have had long-term influences on ecological processes and species composition can have negative cascading effects on certain wild species.

As another example of the impacts of removing pastoralists from a mixed wild and domestic grazing herbivore assemblage, in Tanzania's Ngorongoro Crater, Maasai pastoralists were evicted from living on the caldera floor in 1974 (Homewood and Rodgers 1991). Following the removal of livestock as a permanent presence in the crater (livestock are still allowed to descend into the crater to access mineral licks), different species of ungulates showed both increasing and decreasing trends over the course of the following 20 years (Table 1). Overall, wildlife biomass was unchanged after the removal of pastoralists and their livestock from the Ngorongoro Crater, while wild herbivore diversity declined (Runyoro et al. 1995).

Table 1: Changes in population trends and overall numbers for large mammal herbivores (grazers only) in Ngorongoro Crater following eviction of resident pastoralists in 1974

Species	**Trend before 1974**	**Trend after 1974**	**Overall difference in population before and after 1974**
Wildebeest	Increase	Decline	Decline
Buffalo	Slight increase	Sharp increase	Increase
Hartebeest	Increase	Decline	Increase
Thomson's gazelle	Increase	Decline	Decline
Eland	Constant	Decline	Decline
Zebra	Constant	Decline	Decline

Adapted from Runyoro et al. (1995).

Nelson

Nelson *Pastoralism: Research, Policy and Practice* 2012 2:15 doi:10.1186/2041-7136-2-15

The Loliondo area is a part of the Serengeti ecosystem but, unlike the adjacent Serengeti National Park (SNP) or Ngorongoro Conservation Area (NCA), Loliondo comprises village lands managed by Village Councils. This gives local pastoralist communities a much greater jurisdictional control over land use decisions. Maddox (2003) carried out surveys of both herbivore and large carnivore populations in Loliondo, comparing species densities and abundance to areas inside NCA and SNP. He found no difference in wildlife diversity between the three areas, with animal biomass higher outside SNP, and concludes that there is 'no evidence that the presence of the Maasai and their livestock was particularly detrimental to any single species' and 'no evidence that livestock biomass replaces wild species biomass outside the park.' His findings echo earlier Serengeti ecosystem surveys analysed by Campbell and Borner (1995) which found 'no marked changes in density on either side of the protected area boundaries' dividing community-managed pastoralist lands in Loliondo from the state-managed SNP.

In summary, East African savannahs feature a combination of facilitative and competitive interactions between pastoralists and wildlife, but considerable evidence points to the ecological viability of co-existence between pastoralism and diverse and abundant populations of large mammals. Some studies find few negative spatial influences of livestock on wild ungulate distributions (e.g. Sitters et al. 2009), while other studies find a higher level of competition for forage or water in certain contexts (e.g. Ogutu et al. 2010). As noted above, both the presence and absence of livestock and pastoralists shapes East African rangelands in important ways and thus both positive and negative wildlife conservation influences arise from excluding pastoralists from areas such as national parks (Western and Gichohi 1994). One point on which there is no disagreement, though, is that pastoralist land uses are largely compatible with wildlife, whereas agriculture is not, and the preferences amongst local pastoralist communities for maintaining rangelands as common property has been and remains critical to the integrity of East African savannah ecosystems.

Pastoralist Land Management and Wildlife Conservation in Northern Tanzania

Throughout the East African savannahs of Kenya and Tanzania, wildlife relies extensively on pastoralist lands, both those immediately adjacent to state-protected areas and across the broader landscape. Even in the Serengeti ecosystem, which includes over 25,000 km^2 of state-protected areas, migrating wildlife relies extensively on community lands to the east, west, and north of the park (Thirgood et al. 2004). For example, Norton-Griffiths (1995) estimates that if the ecosystem's migratory wildebeest - roughly 1.2 million animals - could not access the private and communal lands to the north and east of Kenya's Maasai Mara National Reserve, which is the dry season refuge for these herds, the total number of animals the Serengeti ecosystem could support might be reduced by one-third. Indeed, large-scale wildlife declines of resident wildlife populations and the Loita-Mara wildebeest population of 56% and 81%, respectively, have been attributed to land use changes on these private and communal lands (Homewood et al. 2001). In total, 65% of Kenya's wildlife is today found on community and private lands, with only 10% in national parks and about 25% accounted for by the Maasai Mara National Reserve alone (Western et al. 2009a). In other words, pastoralists and other landholders host the majority of Kenya's wildlife populations, sustaining the key asset in the national tourism industry (Norton-Griffiths 2007).

In northern Tanzania, wildlife similarly relies on dispersal areas and habitats outside state-protected areas although, in general, state-protected areas cover larger areas than in Kenya, with Tanzania having devoted about 30% of its total land area to protected areas where human residence is excluded (Brockington 2006).

Northern Tanzania contains two distinct major migratory wildlife systems, with some interaction and movement between the two (Figure 3). The largest number of wildlife lives within the Serengeti ecosystem, which as noted above extends to Kenya's Maasai Mara National Reserve and surrounding private lands. The Serengeti ecosystem is defined by the annual migration of some two to three million wildebeest, zebra, gazelles, and antelopes between dry season habitat in Kenya and the wet season grazing habitat of the Serengeti plains (Sinclair and Arcese 1995). Those plains lie along the border of the SNP and NCA.

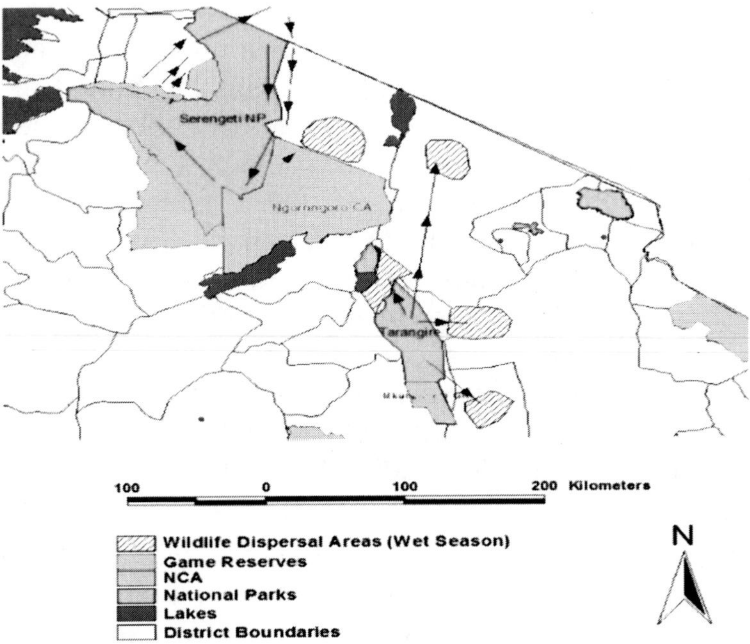

Figure 3: Migration routes and dispersal areas in protected areas and village lands in northern Tanzania's main wildlife areas. Adapted from TNRF (2008).

When the herds of wildebeest move south from the Maasai Mara at the onset of the rains, usually occurring around November, they pass through the community lands of Loliondo. From the Serengeti plains, the herds also spill out onto the Salei plains north of the Gol Mountains, located on the village lands of Malambo and Engare Sero villages. These community-managed areas provide important wet season habitat (Thirgood et al. 2004). To the west of SNP, wildlife also moves outside the boundaries of SNP and the adjacent Ikorongo and Grumeti Game Reserves, onto village lands (now grouped together as the Ikona Wildlife Management Area) in Serengeti and Tarime Districts. Communities resident in these areas are however not transhumant pastoralists but agro-pastoralist members of the Kuria and Ikoma ethnic groups. These communities also prey extensively on wild ungulates as a source of food, in contrast to pastoralists in Loliondo, who do not traditionally eat wild animals. Consequently, wildlife densities to the west of the SNP show a marked decline outside the protected area's confines (Campbell and Borner 1995).

The second extensive wildlife migratory system in northern Tanzania is the Maasai steppe, which covers a vast stretch of about 35,000 km^2, from near the Kenyan border around Lake Natron south through the Simanjiro plains and Tarangire National Park, to the arid Makame depression of northern Kiteto District (Sachedina and Nelson 2010). Unlike the Serengeti system to the west, the Maasai steppe is over 80% community and privately managed lands, with the vast majority being communal village lands. As the name of the ecosystem suggests, the Maasai steppe is inhabited almost entirely by pastoralist communities. The southern part of the steppe borders onto lands occupied by agricultural communities belonging to various non-pastoralist groups; in recent years, these groups of people have been extending smallholder agriculture into traditionally pastoralist areas in the southern Maasai steppe, leading to local conflicts over land use and tenure.

The most important dry season refuge for wildlife of the Maasai steppe is located within Tarangire National Park, which contains the Tarangire River and the Silalo swamps, providing key water sources during droughts to water-dependent species such as elephant, buffalo, and zebra. During the wet season, these and other species disperse widely across the ecosystem (Borner 1985). Short grass plains to the west of Mount Kitumbeine and the Simanjiro plains to the east of Tarangire National Park both provide major dispersal habitats for zebra and wildebeest. In the 1970s, Kahurananga (1981) estimated the wet season populations of the Simanjiro plains at approximately

6,000 zebra and 10,000 wildebeest. Elephants move north out of Tarangire towards Lake Manyara and another small volcanic mountain, Esimingori, during the rains (TCP Tarangire Conservation Project 1997). Another subpopulation of elephants takes refuge in the southern part of Tarangire during droughts but primarily inhabits the Makame depression in the southern Maasai steppe. During the wet season, most wildlife is widely dispersed across the system, with few animals inside Tarangire National Park at that time. Although Tarangire National Park is the main site for large-scale tourism visitation in the system, as one of the leading destinations within the northern tourism circuit, wildlife is largely dependent on habitats located on community lands. The continued maintenance of these wet season habitats and migration corridors between key dry and wet season territories is consequently critical for wildlife and Tarangire's tourism product (Sachedina 2008).

These community lands are managed by local pastoralists for a range of livelihood purposes, but the overwhelming land management aim of pastoralist communities in both the Serengeti ecosystem and the Maasai steppe is oriented towards livestock production. As noted earlier, Maasai pastoralists in northern Tanzania manage their lands according to rules designating different pastures as wet season and as dry season grazing areas. Dry season reserved areas are particularly important in providing a 'grass bank' for livestock to consume during the long dry season of roughly June to October, when forage invariably becomes scarce and animals are stressed for water and nutrients. These dry season reserves are thus a key element of pastoralist livelihoods and natural resource management in northern Tanzania (UCRT 2010).

Traditionally, the use of dry season grazing reserves is governed by decisions made by customary Maasai leadership structures, which involve the leaders (*ilegwanak*) of the different age-sets meeting to decide when community members' herds may access the reserves. The period of time when entry will be allowed will depend on the amount of rain that has fallen in a given year and the condition of forage in the various grazing areas. In recent years, pastoralist communities have been able to formalize their customary land management systems through written village by-laws and land use plans. These measures, prepared based on Tanzania's local government and land policy and legislation, enable communities to formalize their traditional rules and translate them into a modern legalistic form which can be understood by outsiders and given some level of legal force. These locally developed land use plans and by-laws, underpinned by Tanzania's village-based system of local governance and land tenure, also provide the foundations for community-based natural resource management in Tanzania. This includes development of Village Land Forest Reserves under the 2002 Forest Act and Wildlife Management Areas (WMAs) defined by the 2009 Wildlife Conservation Act, as various mechanisms for communities to be granted user rights over these natural resources by the relevant state authorities.

Dry season grazing reserves compose a variable but significant proportion of the communal village lands throughout northern Tanzania's pastoralist areas. These grazing reserves function as *de facto*village-level conserved areas or, to use the IUCN (World Conservation Union) terminology, as Community Conserved Areas (Kothari 2006). The maintenance of reserved grazing areas in pastoralist villages, managed at the communal scale, is a key reason why extensive unfragmented savannah landscapes continue to exist in northern Tanzania, across vast areas with little formal governmental regulation of land uses such as the Maasai steppe. Traditional pastoralist land use practices and designations give northern Tanzania much of its wild or wilderness character, which in turn is a central reason why this landscape both supports wildlife and attracts foreign tourists to the degree that it does.

These grazing reserves also physically overlap with many of the key wildlife habitats and dispersal areas in northern Tanzania. For example, in Loliondo, dry season grazing areas are situated in the western portion of the seven villages whose lands lie adjacent to SNP's eastern boundary (Ngoitiko et al. 2010). Pastoralists' dry season grazing reserves therefore provide a *de facto* buffer zone under local management which helps to conserve wildlife populations in western Loliondo. This includes the resident wildlife populations noted by Maddox (Maddox and Thomas 2003) and others, as well as the southbound migration route of the ecosystem's large ungulate herds. To the east, when wildebeest move onto the Salei plains during the wet season, they utilize dry season grazing reserved areas in Engare Sero village, which straddles the Rift Valley escarpment from Salei down to Lake Natron.

In the Maasai steppe, key habitats such as the Simanjiro plains are partially protected by traditional dry season grazing reserves. For example, Terrat village, which holds about one-third of the Simanjiro plains within its village land boundaries, traditionally allows livestock onto the plains from July up until around January (Nelson et al. 2010). In January, if rain has fallen, wildebeest usually begin calving on the plains, and livestock move away until the dry season grazing period begins again in July (approximately). To the west and southwest, Loiborserrit and Emboreet villages situate their dry season grazing reserves along the eastern border of Tarangire National Park, in an area that provides extensive habitat for buffalo, elephant, and other large mammals moving back and forth across the unseen park boundary. As in Loliondo, these local customary land management practices serve to conserve a large expanse of wildlife habitat adjacent to the park, effectively extending the area conserved and preventing the encroachment of agriculture against the park boundary. During the last 20 years, these traditional designations have been reinforced by a range of formal local governance and conservation measures, including local land use plans and by-laws, community tourism concession agreements with private operators, and, in Terrat village, a unique land easement agreement which pays the village to maintain their portion of the Simanjiro plains as dry season grazing reserve and to formally prohibit agriculture (Sachedina and Nelson 2010).

Estimating the Economic Value of Pastoralist Land Use Practices to Tourism

Pastoralist land management practices across northern Tanzania function to conserve important wildlife habitats, complementing state-protected areas and enabling wildlife to continue moving between seasonal habitats across expansive ecosystems. These customary land use and management practices clearly contribute to the conservation of wildlife habitat and wildlife populations in northern Tanzanian ecosystems. Moreover, these ecosystem services have an important economic dimension, given the scale and importance of northern Tanzania's tourism industry.

Tanzania's tourism industry generated an estimated $900 million in direct revenue and an estimated $1.6 billion in indirect economic value in 2008 (Mitchell et al. 2009). The northern safari circuit, which includes Serengeti, Lake Manyara, and Tarangire National Parks and the Ngorongoro Conservation Area (NCA, and adjacent pastoralist lands, attracts about 300,000 international tourists and total in-country tourist expenditure estimated at around $550 million annually (Mitchell et al. 2009). In other words, the northern safari circuit, which relies entirely on wildlife using both state-protected areas and local pastoralists' lands (even if most tourism activity is situated within the former rather than the latter), accounts for over 60% of Tanzania's total national tourism industry.

It is clear that by conserving large proportions of northern Tanzania's wildlife ecosystems, local pastoralist communities collectively make an important contribution to the national and regional economy. But, precisely how much do these ecological services contribute to the overall wildlife-based tourism economy? What proportion of total northern circuit tourism earnings is attributable to pastoralist land management practices?

In answering this question, one faces two basic challenges. First, the quality of available data on tourism revenues and aggregate economic impacts is limited and somewhat dated. This article utilizes the most reliable and informative statistics and analyses, building in particular on the northern circuit tourism value chain analysis produced by Mitchell et al. (2009). Second, the role of pastoralist land use practices varies across individual landscapes within northern Tanzania. In the Maasai steppe, for instance, wildlife that spends the dry season inside Tarangire National Park is dependent through the entire wet season, including for calving grounds, on community lands outside the park. In the Serengeti ecosystem, state-protected areas cover a much greater proportion of the total ecosystem. Thus, the economic contribution of pastoralist land uses across northern Tanzania needs to be weighted according to the relative importance of pastoralist land uses in these different ecosystems anchoring the regional tourism industry.

Using the total number of tourist visitors to the different areas (NCA and SNP for the Serengeti ecosystem; TNP and LMNP for the Maasai steppe ecosystem), the proportional values of the two systems in relation to the northern circuit's total direct economic value of $550 million can be estimated.[a] Dividing this total between the two ecosystems that comprise the northern circuit, based on visitor numbers at the different areas, results in a total value attributable to the Maasai steppe of about $153 million, and for the Serengeti ecosystem (including 25% of NCA revenue) of $224 million (Table 2).[b]

[a]This figure is taken from Mitchell et al. (2009) and represents

the total revenue expenditure by the approximately 300,000 foreign visitors to the northern circuit.

[b]NCA includes both a large portion of the Serengeti plains, which are a part of the greater Serengeti ecosystem, and the relatively enclosed Ngorongoro Crater, the primary tourism site within NCA, where wildlife is generally resident and does follow the migratory patterns of wildlife elsewhere in the Serengeti system. If we only include 25% of the total revenue from NCA, then we obtain a perhaps more realistic figure for the Serengeti ecosystem that excludes Ngorongoro Crater of about $224 million.

Table 2: Proportional value of different areas (national parks and NCA) in the northern circuit

Area	Number of tourists	Proportion of northern circuit visitors (%)	Estimated proportion of total northern circuit revenue generation
Lake Manyara NP	141,226	16	$88,957,628
Serengeti NP	265,880	31	$167,476,628
Tarangire NP	102,562	12	$64,603,347
NCA	360,000	41	$226,762,397
Total value attributable to Maasai steppe			$153,560,975
Total value attributable to Serengeti ecosystem (including 25% NCA)			$224,167,227

This is based on the proportion of northern circuit visitor numbers at each area. Adapted from TANAPA (2007) and Mitchell et al. (2009).

Nelson

Nelson *Pastoralism: Research, Policy and Practice* 2012 2:15 doi:10.1186/2041-7136-2-15

In the Maasai steppe, over 80% of the total wildlife range lies on private and communal lands, the vast majority of it within predominantly pastoralist villages. However, the key wildlife habitats comprise a smaller area and are focused on areas such as the Simanjiro plains and similar short grass plains near Mount Kitumbeine for herds of zebra and wildebeest, as well as areas such as Lolkisale and Makame for elephants. We can conservatively estimate that about 50% of the key habitats supporting wildlife in the Maasai steppe is managed by pastoralists on village lands in this ecosystem and that, if all pastoralist lands were unavailable for use by wildlife, the total value of the ecosystem in terms of wildlife-based tourism would decline by half.

In the Serengeti ecosystem, the proportion of wildlife habitat provided by pastoralist communities' village lands is much lower. Thirgood et al. (2004), using radio-telemetry studies of wildebeest during 1999 and 2000, estimate that the wildebeest herds spend about 3% of their total time in Loliondo and the Salei plains, which are the main pastoralist-controlled lands in the Serengeti ecosystem.[c] We will therefore estimate the relative contribution that pastoralists make to the conservation of the Serengeti ecosystem at 3%.[d]

[c]Although non-pastoralist community lands to the west of SNP, in the Ikoma area, and to the north of the Maasai Mara National Reserve in Kenya are more important to wildebeest and other migratory wildlife, these areas are not included in the scope of this study, which is concerned only with pastoralist lands in Tanzania.

[d]This figure is also conservative because it does not in any way factor in the role pastoralists have played for hundreds of years in maintaining the landscape of the NCA, which, unlike the national parks is a multiple land use area with over 50,000 resident Maasai pastoralist residents (Homewood and Rodgers 1991). However, since pastoralists in this area, unlike in Loliondo, do not have clear rights over land management decisions and are not able to make and enforce their own land use plans due to regulatory powers exercised by the Ngorongoro Conservation Area Authority (Shivji et al. 1998), it is difficult to credit pastoralists in NCA with responsibility for land management decisions that contribute to conservation in the way that one may in other pastoralist areas.

The approximate figures of 50% of habitat in the Maasai steppe, and 3% in the Serengeti, as representing the degree to which wildlife in these ecosystems depend on pastoralist lands adjacent to the state-protected areas enables one to attach valuation figures, based on the total revenue generated through tourism, to pastoralist lands in these areas. This leads to an estimate of the total economic value in relation to wildlife conservation and tourism revenue of approximately $83.5 million annually (Table 3). This figure is somewhat imprecise, and uses a cobbled-together methodology in light of limitations in the available data, but nevertheless demonstrates that the economic contribution that pastoralists make through their land management practices in relation to Tanzania's wildlife-based tourism industry is substantial.

Table 3: Estimated value of pastoralist lands to wildlife-based tourism in the Maasai steppe and Serengeti ecosystems

Ecosystem	Degree of wildlife reliance on pastoralist lands (%)	Economic contribution of pastoralist lands as a percentage of total tourism earnings
Maasai steppe	50	\$76,780,487.73
Serengeti ecosystem	3	\$6,725,016.81
Total		\$83,505,504.54

Based on the relative degree that wildlife relies on pastoralist lands outside the boundaries of state-protected areas.

Nelson

Nelson Pastoralism: Research, Policy and Practice 2012 2:15 doi:10.1186/2041-7136-2-15

It should also be noted that this estimate can be considered a minimum valuation of pastoralist contributions to the wildlife-tourism economy. These estimates focus on the northern safari circuit's two main ecosystems - the Serengeti and the Maasai steppe - but other less significant forms of wildlife utilization also generate revenue. For example, most pastoralist lands are allocated by the central government as tourist hunting concessions, which as of 2003 generated an average of \$40/km^2 in direct government revenue (Baldus et al. 2004). Nearly all of such predominantly pastoralist districts such as Longido, Monduli, and Simanjiro are allocated as hunting concessions, covering an area of at least 35,000 km^2. The revenue from tourist hunting in pastoralist areas, which is largely enabled by the co-existence of pastoralist communities with diverse and relatively abundant populations of large mammals on their lands, is therefore at least \$1.4 million annually. Due to recent raises in hunting fees, the current actual total is in all likelihood several times that amount in direct revenue alone, without including any indirect benefits or economic multipliers.

Policy Implications

It has long been recognized that pastoralist lands and land use practices provide much of the habitat that migratory wildlife in East African national parks continues to rely on for seasonal habitat requirements (Myers 1972; Western 1982). As northern Tanzania's tourism industry has grown, with the northern safari circuit comprising about 6% of the country's total national economy, the economic importance of pastoralist conservation practices in these extensive ecosystems has increased considerably. This highlights the strategic economic importance to the Tanzanian state of pastoralist land use practices, including in particular the rights to zone land for seasonal uses and continue to practice mobility as a key element of management systems.

Given the economic importance of pastoralist land management practices, Tanzanian policy in sectors such as land, livestock, wildlife, and tourism should take into account these strategic economic considerations and support the ability of local pastoralist communities to maintain communal pastures and reserved grazing areas. The key, from an economic and development policy perspective, is to craft policy and legal measures that enable pastoralist rangelands to continue to maintain wildlife and livestock in communally managed rangelands. Land use changes that result in conversion of these pastoralist rangelands to agricultural cultivation represent the foremost threat to wildlife populations and, in turn, the continued market advantage Tanzania enjoys in high-end wildlife tourism.

It is thus highly notable, and in an economic policy sense quite perverse, that Tanzanian policy generally provides few incentives for pastoralists to continue to tolerate or conserve wildlife on their lands, and nor are policy measures effectively geared towards encouraging pastoralists to maintain the communal land holdings and mobile transhumant production systems that underpin the continued co-existence of wildlife and livestock in these rangelands.

Land Tenure

A key issue for the future of Tanzanian savannah ecosystems will be the ability of local pastoralist communities to continue managing rangelands at the communal scale, rather than fragmenting rangelands into individually owned properties. Converting rangeland from communal pastures to individual properties will likely prevent the movements across large areas of land that both livestock and wildlife in semi-arid landscapes depend upon. Tanzania needs look no further for a possible future scenario than Kenyan Maasailand to the north, where the former Group Ranches have been extensively subdivided into individual land holdings which are of insufficient size to support pastoralist production (Kimani and Pickard 1998; Western et al. 2009b). These processes of subdivision have been driven by weaknesses in the governance and design of the Group Ranches as collective entities, resulting in progressive loss of communal land through decisions made by local elites on the Group Ranch governing bodies (Mwangi 2007). This process of rangeland fragmentation, driven by past policy design choices by the Kenyan state, has dire implications for wildlife in key areas such as the Maasai Mara and Amboseli - also the country's most valuable wildlife tourism sites - and there is evidence of declining wildlife populations due to these land use changes (Homewood et al. 2001).

Unlike Kenya, Tanzania's land tenure laws and policies provide a relatively enabling framework for communal land management. The Land Act and Village Land Act, passed in 1999, make customary rights in land legally equivalent to granted rights and specifically cater for rights to be held and exercised collectively (Alden and Liz 2003). The Village Land Act also provides for villages to zone their land between individual and communal areas, which supports land use planning measures that reserve different seasonal grazing areas, as has been done in many northern Tanzanian pastoralist communities (UCRT Ujamaa Community Resource Team 2010).

However, the Land Act also includes 'unused community lands' within its definition of 'general lands' which are under the control of the Ministry of Lands and may be allocated for use to applicants. Seasonally used pastoralist lands are often perceived by outsiders as 'unused' and therefore may be subject to attempts by government authorities to allocate them in this way (Mattee and Shem 2006). Lands that pastoralists have traditionally used and set aside as dry season grazing reserves are often considered unused, and recently, during certain land disputes, notably the land conflict that has occurred in Loliondo since 2009, government has argued that such areas are not in fact village lands (TNRF and Maliasili Initiatives 2011). This undermines the security of communal pastoralist land holdings and encourages individualization and cultivation of pastoralist lands, in a process thatSachedina (2008) terms as 'branding the land' with the plough so that use and occupancy can decisively be documented.

Livestock

Tanzanian livestock policy and legal reforms in recent years have consistently emphasized modernization and increasing commercial meat, milk, and hide production. These modernization models generally prioritize development of large-scale, privately owned ranches as the key to increasing commercial production and view traditional pastoralist production as inefficient and relatively archaic (Mattee and Shem 2006).

The individualization or spread of privately owned ranches in northern Tanzania would have significant costs in relation to the tourism industry, due to the impacts on wildlife of rangeland fragmentation. When private ranches are carved out of communal pastures, the private parcels tend to be much smaller and are often demarcated through fencing. It is debatable if the development of smaller, privately held landholdings out of larger scale, communal pastures would lead to more commercial production of livestock in northern Tanzania. However, it is indisputable that such changes would have major costs in terms of wildlife populations and tourism enterprises, as Kenya's experience with rangeland fragmentation illustrates (Norton-Griffiths 2007). Rather than promoting private ranch-based livestock sector development models, policymakers should focus on measures that enable pastoralist livestock producers to improve their production and profit margins, such as enhanced market information or veterinary services.

Wildlife and Tourism

While pastoralists continue to provide valuable ecosystem services in support of Tanzania's tourism industry and related wildlife conservation interests, the wildlife and tourism sectors continue their long legacy of failing to reciprocate in their provisions as they relate to pastoralists. Historically, pastoralists have lost key lands and resources to the state-protected areas and continue to lose access to areas. For example, the Mkungunero Game Reserve's creation in the late 1990s in the southern Maasai steppe created conflicts with local pastoralist villages and widespread apprehension about expanding protected areas and land loss in the area (Sachedina 2008). The allocation of centrally managed tourist hunting concessions on community lands is a widespread source of conflict and results in local people having insecure rights over land and resources in areas such as the Maasai steppe (Sachedina 2008). Pastoralists in Loliondo and the Maasai steppe have also had recurrent disputes with protected area authorities regarding the boundaries of their lands and national parks (Goldman 20092011).

These tensions between protected area authorities and local communities create perverse incentives which are not conducive to maintaining integrated wildlife and livestock land uses in pastoralist rangelands. For example, in Emboreet village, on the border of Tarangire National Park, there is evidence that local pastoralists have expanded farming in key wildlife habitats as a response to fears that external agencies will appropriate their land for purposes of wildlife conservation (Sachedina 2008).

While the national park service, TANAPA, does maintain some benefit-sharing programmes with surrounding communities, these are neither of the scale nor of the type needed to positively influence neighbouring pastoralists to support ecosystem-scale conservation objectives through the necessary land use choices. For example, Tarangire National Park provided average benefits (usually in the form of contributions to district or village-level social service or infrastructure projects) to Simanjiro District of $30,470; at this time, the park's annual income from tourism was around $1.5 million/year (reaching $2.1 million by 2005 and now consistently above $2 million/year) (Sachedina and Nelson 2010). This represents an investment - designed explicitly to encourage support for conservation among neighbouring communities - of about 2% of revenues in what is the single most important dispersal and breeding area for much of the wildlife found in Tarangire. As Sachedina and Nelson (2010) describe, it is not only the limited amount of funds invested in neighbouring communities, but the way these funds are provided as in effect unconditional donations to development projects that are not in any way tied to local land use decisions or conservation actions, that undermines their impact. By contrast, an investment by a coalition of private tourism operators of a limited amount of funding in a conditional agreement with one village on the Simanjiro plains which agreed not to farm within its traditional grazing area has reinforced local incentives not to convert key wildlife habitat to agriculture (Nelson et al. 2010). In order to protect its own long-term financial interests in wildlife and tourism, it would be advisable for TANAPA to make more targeted investments in supporting pastoralist land uses in areas surrounding its parks and to use conditional payment for ecosystem services agreements to achieve such ends.

At the level of wider wildlife and tourism policy, recent years have witnessed a perverse trend towards increasing restrictions on the ability of local communities, such as pastoralists living adjacent to Tarangire and SNPs, to capture revenue directly from concessionary agreements with private tourism operators. In 2007, the government introduced new regulations for fee-sharing systems for tourism occurring on community lands, effectively taxing these agreements at a punitive level in excess of 50%, thus greatly reducing revenues that communities receive from such arrangements (Snyder et al. 2011).

At the same time, Tanzania has moved forward with the development of WMAs, areas set aside on village lands by communities where the government grants communities formal user rights over wildlife so that they can capture benefits from commercial uses such as tourism and hunting. However, the development of WMAs has been limited by complex procedural requirements, limited investments in local capacity building, and benefit-sharing arrangements that still give government 35% of total revenues and do not devolve control over tourist hunting to the local level (Nelson 2007; Sulle et al. 2011). In some areas such as Loliondo, pastoralists have opted not to implement proposed WMAs due to concerns about these factors as well as the security of grazing areas under the wildlife-focused WMA framework (Ngoitiko et al. 2010). In theory, WMAs should be able to reconcile local livelihoods and pastoralist land use objectives with wildlife conservation, particularly given the historic co-existence of wildlife with pastoralist land use systems. It would be economically rational, from the perspective of the state's interests in the larger wildlife systems that these WMAs form a part of, for the state to give communities 100% of revenues from WMAs[e] and for the state to facilitate greater opportunities from tourism and tourist hunting to generate more revenue and ensure WMAs are competitive forms of land use. This would enhance the likelihood that pastoralist communities embrace the WMA option and that WMAs create effective incentives to maintain integrated wildlife and livestock land uses.

[e]For example, in the notionally similar communal conservancies of Namibia, communities are legally entitled to 100% of all revenue generated in the conservancies.

CONCLUSIONS

The ability of transhumant pastoralist livestock production to co-exist alongside a high level of diversity and abundance of wild ungulates and other large mammals in northern Tanzania is not only a key to wildlife conservation in the region but also to the area's role as the driver of Tanzania's tourism industry, which now comprises around 11% of the country's total economic production. Pastoralists in northern Tanzania, through their land use choices and specific management practices such as the maintenance of dry season grazing reserves and protection of forests and woodlands, provide what is in reality an increasingly commercially valued ecosystem service.

Despite the value pastoralist land uses provide to the growing wildlife tourism industry, Tanzanian policies, particularly wildlife policy and legislation, continue to place constraints on pastoralist land uses, particularly in terms of the maintenance of communal land uses and ability of communities to capture revenues from tourism. These policy choices serve to discourage continued collective pastoralist land uses and play a role in the expansion of agricultural cultivation and fragmentation of communal rangelands. Tanzanian policy is inefficient in the sense that it discourages local land use choices that help to support a critical segment of the national economy. A more rational and productive set of policy choices would not only remove disincentives to pastoralist land uses, but also develop mechanisms for actively rewarding them, for example, through payments for ecosystem services that better link revenues from tourism with local communities whose land uses support wildlife and tourism. Tanzanian policymakers, pastoralist communities, wildlife conservation interests, and the tourism industry share a common economic interest in the design and deployment of such mechanisms so that pastoralist land uses will continue to produce the ecosystem services that they presently provide.

ACKNOWLEDGEMENTS

This research was originally carried out with the support from the International Institute for Environment and Development. In particular, Ced Hesse and James MacGregor provided valuable advice and support in the initial design of the study and commented on earlier versions of this article. Two anonymous reviewers also provided helpful comments and suggestions.

REFERENCES

1. Alden, W, and A Liz. 2003. Community-based land tenure management: questions and answers about Tanzania's New Village Land Act. London: International Institute for Environment and Development.
2. Baldus, , D Rolf, and AE Cauldwell. 2004. Tourist hunting and its role in development of wildlife management areas in Tanzania. Paris: Paper presented to the Sixth International Game Ranching Symposium. 6–9 July, 2004
3. Barrow, E. 1988. Trees and pastoralists: the case of the Pokot and Turkana. Overseas Development Institute, Social Forestry Network, Network Paper 6b. London: ODI.
4. Behnke, RH, I Scoones, and C Kerven. 1993. Range ecology at disequilibrium: new models of natural variability and pastoral adaptation in African savannas. London: Overseas Development Institute.
5. Bell, RHV. 1971. A grazing system in the Serengeti. Scientific American 224: 86–93.
6. Berkes, F. 1989. Common property resources: ecology and community-based sustainable development. London: Belhaven Press.
7. Borner, M. 1985. The increasing isolation of Tarangire National Park. Oryx 19: 91–96.
8. Bourn, D, and R Blench. 1999. Can livestock and wildlife co-exist? An interdisciplinary approach. Oxford: Overseas Development Institute.

9. Brockington, D. 2006. The politics and ethnography of environmentalisms in Tanzania. African Affairs 105: 97–116.
10. Campbell, K, and M Borner. 1995. Population trends and distribution of Serengeti herbivores: implications for management. In Serengeti II: dynamics, management, and conservation of an ecosystem, ed. Sinclair ARE, Arcese P, 117–145. Chicago: University of Chicago Press.
11. Collet, D. 1987. Pastoralists and wildlife: image and reality in Kenya Maasailand. In Conservation in Africa: people, policies, and practice, ed. Anderson D, Grove R, 129–148. Cambridge: Cambridge University Press.
12. Dodd, JL. 1994. Desertification and degradation in sub-Saharan Africa: the role of livestock. Bioscience 44 (1): 28–34.
13. Du Toit, J. 2011. Coexisting with cattle. Science 333: 1710–1711.
14. Ellis, JE, M David, and Swift. 1988. Stability of African pastoral ecosystems: alternate paradigms and implications for development. Journal of Range Management 41 (6): 450–459.
15. Gichohi, H. 1992. The effects of fire and grazing on grasslands of Nairobi National Park. MSc Thesis. Nairobi: University of Nairobi.
16. Goldman, M. 2009. Constructing connectivity? Conservation corridors and conservation politics in East African rangelands. Annals of the Association of American Geographers 99 (2): 335–359.
17. Goldman, M. 2011. Strangers in their own land: Maasai and wildlife conservation in Northern Tanzania. Conservation and Society 9 (1): 65–79.
18. Grzimek, B, and M Grzimek. 1960. Serengeti shall not die. New York: E.P. Dutton & Co., Inc.
19. Hardin, G. 1968. The tragedy of the commons. Science 162: 1243–1248.
20. Hayes, TM. 2006. Parks, people, and forest protection: an institutional assessment of the effectiveness of protected areas. World Development 34: 2064–2075.
21. Homewood, KM, and WA Rodgers. 1991. Maasailand ecology: pastoralist development and wildlife conservation in Ngorongoro, Tanzania. Cambridge: Cambridge University Press.

22. Homewood, K, EF Lambin, E Coast, A Kariuki, I Kikula, J Kivelia, M Said, S Serneels, and M Thompson. 2001. Long-term changes in Serengeti-Mara wildebeest and land cover: pastoralism, population, or policies? PNAS 98 (22): 12544–12549.
23. Kahurananga, J. 1981. Population estimates, densities and biomass of large herbivores in Simanjiro Plains, northern Tanzania. African Journal of Ecology 19: 225–238.
24. Kimani, K, and J Pickard. 1998. Recent trends and implications of group ranch subdivision and fragmentation in Kajiado District, Kenya. The Geographical Journal 164: 202–213.
25. Kothari, A. 2006. Community conserved areas: towards ecological and livelihood security. Parks 16: 3–13.
26. Lamprey, H. 1983. Pastoralism yesterday and today: the overgrazing problem. In Tropical Savannas: ecosystems of the world, ed. Bourliere F, 643–666. Amsterdam: Elsevier.
27. Maddox, , and M Thomas. 2003. The ecology of cheetahs and other large carnivores in a pastoralist-dominated buffer zone. PhD Thesis. London: University College, London and Institute of Zoology.
28. Mattee, AZ, and M Shem. 2006. Ambivalence and contradiction: a review of the policy environment in Tanzania in relation to pastoralism. Drylands Issue Paper No. 140. London: International Institute for Environment and Development.
29. McCabe, JT. 1990. Turkana pastoralism: a case against the Tragedy of the Commons. Human Ecology 18 (1): 81–103.
30. Mitchell, J, K Jodie, and L Jenny. 2009. Making success work for the poor: Package tourism in northern Tanzania. Overseas Development Institute and SNV. http://www.odi.org.uk/resources/docs/4203.pdf . Accessed. 14/7/2012
31. Mwangi, E. 2007. The puzzle of group ranch subdivision in Kenya's Maasailand. Development and Change 38 (5): 889–910.
32. Myers, N. 1972. National parks in savannah Africa. Science 178: 1255–1263. |
33. Nelson, F. 2007. Emergent or illusory? Community wildlife management in Tanzania. Drylands Issue Paper No. 146. London: International Institute for Environment and Development.

34. Nelson, F, C Foley, S Lara, AL Foley, E Loure, D Peterson, M Peterson, T Peterson, H Sachedina, and A Williams. 2010. Payments for ecosystem services as a framework for community-based conservation in northern Tanzania. Conservation Biology 24 (1): 78–85.
35. Neumann, RP. 1998. Imposing wilderness: struggles over livelihood and nature preservation in Africa. Berkeley: University of California Press.
36. Ngoitiko, M, M Sinandei, P Meitaya, and F Nelson. 2010. Pastoral activists: negotiating power imbalances in the Tanzanian Serengeti. In Community rights, conservation and contested land: the politics of natural resource governance in Africa, ed. Nelson F, 269–289. London: Earthscan.
37. Norton-Griffiths, M. 1995. Economic incentives to develop the rangelands of the Serengeti: implications for wildlife conservation. In Serengeti II: dynamics, management, and conservation of an ecosystem, ed. Sinclair ARE, Arcese P, 588–604. Chicago: University of Chicago Press.
38. Norton-Griffiths, M. 2007. How many wildebeest do you need? World Economics 8: 41–64.
39. Odadi, , O Wilfred, MK Karachi, SA Abdulrazak, and TP Young. 2011. African wild ungulates compete with or facilitate cattle depending on season. Science 3 333: 1753–1755.
40. Ogutu, JO, H-P Piepho, S Robin, ME Reid, RL Rainy, JS Kruska, MN Worden, and N Thompson Hobbs.. 2010. Large herbivore responses to water and settlements in savannas. Ecological Monographs 80 (2): 241–266.
41. Ostrom, E. 1990. Governing the commons: the evolution of institutions for collective action. Cambridge: Cambridge University Press.
42. Ostrom, E, and H Nagendra. 2006. Insights on linking forests, trees, and people from the air, on the ground, and in the laboratory. PNAS 103: 19224–19231.
43. Pratt, DJ, and MD Gwynne. 1977. Rangeland management and ecology in East Africa. London: Hodder and Stoughton.
44. Runyoro, , A Victor, H Heribert, EB Chausi, and PD Moehlman. 1995. Long-term trends in the herbivore populations of the

Ngorongoro Crater, Tanzania. In Serengeti II: dynamics, management, and conservation of an ecosystem, ed. Sinclair ARE, Arcese P, 146–168. Chicago: University of Chicago Press.

45. Sachedina, H. 2008. Wildlife is our oil: conservation, livelihoods and NGOs in the Tarangire ecosystem, Tanzania. UK: PhD thesis, University of Oxford.
46. Sachedina, H, and F Nelson. 2010. Protected areas and community incentives in savannah ecosystems: a case study of Tanzania's Maasai Steppe. Oryx 44: 390–398.
47. Scoones, I. 1995. Living with uncertainty: new directions in pastoral development in Africa. London: Intermediate Technology Publications.
48. Shivji, , G Issa, and WB Kapinga. 1998. Maasai rights in Ngorongoro, Tanzania. London: IIED/HAKIARDHI.
49. Sinclair, ARE, and P Arcese. 1995. Serengeti II: dynamics, management, and conservation of an ecosystem. Chicago: University of Chicago Press.
50. Sitters, J, H Ignas MA, H Milena, and O Gordon SO. 2009. Herded cattle and wild grazers partition water but share forage resources during dry years in East African savannas. Biological Conservation 142: 738–750.
51. Snyder, , A Katherine, and EB Sulle. 2011. Tourism in Maasai communities: a chance to improve livelihoods? Journal of Sustainable Tourism 19 (8): 935–951.
52. Sulle, E, E Lektita, and F Nelson. 2011. From promise to performance? Wildlife Management Areas in Northern Tanzania. Arusha: Tanzania Natural Resource Forum, Ujamaa Community Resource Team and Maliasili Initiatives.
53. TANAPA (Tanzania National Parks). 2007. Annual report and financial statements 2006/07. Arusha: TANAPA.
54. TCP (Tarangire Conservation Project). 1997. Analysis of migratory movements of large mammals and their interaction with human activities in the Tarangire Area in Tanzania as a contribution to a conservation and sustainable development strategy. Final Report: Explanatory Section, Arusha.
55. Thirgood, S, M Anna, T Sebastian, H Grant, M Ephraim, M Titus, K Morris, F John, S Anthony RE, and B Markus. 2004. Can parks

protect migratory ungulates? The case of the Serengeti wildebeest. Animal Conservation 7: 113–120.

56. TNRF (Tanzania Natural Resource Forum). 2008. Wildlife for all Tanzanians: stopping the loss, nurturing the resource and widening the benefits. An information pack and policy recommendations. Arusha: TNRF.

57. TNRF and Maliasili Initiatives. 2011. Integrating pastoralist livelihoods and wildlife conservation? Options for land use and conflict resolution in Loliondo division, Ngorongoro District. Arusha: Tanzania Natural Resource Forum.

58. UCRT (Ujamaa Community Resource Team). 2010. Participatory land use planning as a tool for community empowerment in northern Tanzania. Gatekeeper Series No. 147. London: International Institute for Environment and Development.

59. Western, D. 1982. Amboseli National Park: enlisting landowners to conserve migratory wildlife. Ambio 11 (5): 302–308.

60. Western, D. 1989. Conservation without parks: wildlife in the rural landscape. In Conservation for the twenty-first century, ed. Western D, Pearl M, 158–165. Oxford: Oxford University Press.

61. Western, D, and H Gichohi. 1994. Segregation effects and the impoverishment of savannah park: the case for ecosystem viability analysis. African Journal of Ecology 31: 269–281.

62. Western, D, S Russell, and I Cuthill. 2009. The status of wildlife in protected areas compared to non-protected areas of Kenya. PLoS ONE 4 (7): e6140. | |

63. Western, D, R Groom, and J Worden. 2009. The impact of subdivision and sedentarization of pastoral lands on wildlife in an African savanna ecosystem. Biological Conservation 142: 2538–2546.

Chapter 8

Livestock Mobility and Animal Health Policy in Southern Africa: the Impact of Veterinary Cordon Fences on Pastoralists

Daniel J McGahey

African Wildlife Foundation, Conservation Centre, Ngong Road, Karen, 00502, Nairobi, Kenya

ABSTRACT

Greater understanding of how policies are restricting extensive pastoral production is needed, given the importance of livestock mobility to both social and environmental health in drylands. In southern Africa, communal pastoral rangelands continue to be enclosed and dissected by large-scale barrier fences designed to control livestock diseases and protect lucrative livestock export agreements. This paper looks at how these veterinary cordon fences are changing pastoral systems in Botswana, particularly patterns of resource access and livestock mobility. The results show how overall consensus of opinion towards fencing among pastoralists is positive; fences do restrict resource access yet on balance herding demands are reduced and livestock security improved. However, a minority of marginalised groups recognise that fencing restricts risk management strategies and increases competition for key resources, raising concern over more widespread social differentiation and increased vulnerability at the regional level. The paper demonstrates why a closer examination of the trade-offs and consequences associated with animal health policy options in Africa is needed, particularly to determine policy-marketing pathways that allow more flexible and opportunistic resource access within communal pastoral areas.

BACKGROUND

'Clearly, there will be occasions when movement restrictions to limit the spread of contagious diseases and the imposition of quarantine regulations are warranted. However, in the design of veterinary regulations and associated administrative arrangements, the trade-off between veterinary control and mobility must be taken into account' (Scoones 1995:17)

Livestock mobility is the principal means by which pastoralists cope with and take full advantage of natural resource variability in drylands (Scoones 1995, Behnke et al. 1993). Over the last two decades, scholars working with pastoral groups from both the northern and southern hemisphere have reported an almost universal decline in herd mobility, caused by a variety of factors ranging from the commoditisation of the pastoral economy to villagisation policies (cf. Fratkin 1997). Sedentarisation has been particularly influenced by declines in unfenced land available to pastoralists and their livestock after the widespread privatisation and enclosure of pastures for ranching, mechanised agriculture or conservation. Much of this change has been a consequence of the negative policy discourse surrounding pastoralism that often manifests in severely repressive rangeland policies designed to reform practices seen as irrational or environmentally destructive, whilst simultaneously improving economic productivity.

Formulating legislation that supports the spatial and temporal dynamism of natural resource use is now seen by many as one of the most important challenges for pastoral drylands (UNDP 2003,Niamir-Fuller

1999, Fratkin and Mearns 2003)[a] . This follows the wider recognition of a 'new paradigm' for understanding dryland pastoral systems drawing heavily upon concepts of ecosystem instability or 'disequilibrium ecological theory' (Behnke et al. 1993). Advocates of this approach have shown how the policy implications are profound, requiring major changes in thinking at both the state and local level (Niamir-Fuller and Turner 1999, Scoones 1995). The future sustainability of pastoral drylands may well be dependent on the ability of decision makers to support herd mobility within almost all aspects of governance and service provision in drylands (Fratkin and Mearns 2003).

Footnote

[a]Globally, drylands occupy 41% of the global land surface and are home to 35% of the population (Safriel et al. 2005). Of these, an estimated 100 to 200 million are pastoralists and the production system is the most widespread on earth practised on at least two-thirds of global drylands (Davies and Hatfield 2007). Throughout this paper, I use the term pastoral drylands to refer to development issues present in these regions.

Recognition of the importance of herd mobility in determining environmental sustainability in non-equilibrium pastoral drylands has prompted researchers to understand drivers determining its contemporary use (Baker and Hoffman 2006, Coppolillo 2000, Adriansen and Nielsen 2002, Turner and Hiernaux 2002) and the ecological consequences of its decline (Reid et al. 2003, Boone and Hobbs 2003). Many of these studies have reported the growing importance of socio-economic rather than environmental factors in determining herding decisions at a microscale by influencing the availability of herding labour, quality of herd management and daily herd itineraries (cf. Baker and Hoffman 2006, Turner 1999). Most recent work has demonstrated how, even in industrialised countries, the use of livestock mobility remains an important strategy in response to environmental variability (McAllister 2010, Manzano and Casas 2010).

One of the main issues restricting livestock movement raised early in the 'new rangeland ecology' literature was the abundance of colonial-era policies founded upon conventional stability concepts. One specific area of policy, animal health, represents a conspicuous obstacle to pastoral mobility because the effective control of livestock diseases under most disease control legislation in Africa requires strict livestock movement restrictions. In discussing animal health regulations in relation to mobility Scoones (1995:16-17) noted for example:

'Most administrative arrangements (movement permits, veterinary regulations, etc.) assume a stable environment where movement is discouraged. However, under more dynamic ecological conditions, movement becomes increasingly central and such administrative structures impose a major cost on the production system.'

Until relatively recently, little specific research has sought to understand exactly what cost policies such as animal health legislation have on the pastoral production system and how these might be reduced. Notable exceptions include recent work on livestock marketing in eastern Africa that has begun to demonstrate the impediments colonial-era animal health legislation designed for state-led marketing has on the growth of regional private sector-led trade, for example (McPeak and Little 2006).

In southern Africa, sustained public sector investment in these disease control policies has facilitated the attainment of 'disease freedom' from certain areas, primarily through the construction of a network of large-scale veterinary cordon fences (Scoones et al. 2010). These facilitate the control of livestock movements upon extensive rangelands and allow the area-based disease freedom required to access lucrative beef export markets (Sutmoller 2002). By restricting mobility and dividing communal rangelands, veterinary cordon fences could have potentially detrimental impacts upon pastoralists and their flexible systems of resource access, yet apart from several brief time-bound assessments the exact societal costs of this approach remain poorly understood (see, e.g. Hobbs 1981, Hitchcock 1995, 2002, Scott Wilson Resource Consultants 2000).

To address this knowledge gap, I investigated the impact of veterinary cordon fences upon livelihoods and natural resource access for pastoralists of Chukumuchu, a small settlement located in a heavily fenced area of Ngamiland district, northwest Botswana. The following section briefly describes the study area and situates it within Botswana's expanding network of veterinary cordon fences. Next, the paper provides a short review of the methods used in the research before comparing existing patterns of pastoral management, settlement and mobility with those recorded in the literature, primarily by anthropologists working in the 1970s and 1980s. This shows that mobility has declined and the remaining sections describe and analyse how pastoral society and natural resource management has become progressively more sedentary for a variety of reasons. The influence of the region's veterinary cordon fences upon pastoral mobility and resource use is then extracted from this narrative of sedentarisation.

The described changes demonstrate how enclosure has increased societal vulnerability to the various risks associated with sedentary pastoralism in the Kalahari. Moreover, fencing has added to factors promoting the increasing commercialisation of the pastoral sector and disadvantaged younger herders by obstructing many of the traditional pathways into pastoralism. I argue that given these findings significant attention should be placed on current efforts to reform animal health policies in Africa, particularly at how the interests of natural resource-dependent pastoralists seeking flexible and opportunistic resource access arrangements can be reconciled against those of the commercial livestock lobby.

Study Area

Veterinary Cordon Fencing in Botswana

For the purposes of disease control, particularly foot-and-mouth disease which remains endemic in wild buffalo, Botswana is dissected by a network of veterinary cordon fences which divide the land into four: an export zone, buffer zone, vaccination/surveillance zone and a wildlife/foot-and-mouth disease (FMD) zone. Within these zones over 80% of livestock reared for the commercial sector are grazed on communal land (Darkoh and Mbaiwa 2002). The remaining cattle are reared on fenced tribal lands (tribal grazing lands policy, TGLP) or freehold farms, with the privatised land of the latter covering a meagre 6% of the country's total land area (Ndzinge et al. 1984). Elsewhere in southern Africa (Namibia, South Africa and formally Zimbabwe) a dualistic system exists whereby disease control, market infrastructure and support is focussed on the privatised ranching sector enclosed and protected from diseases endemic in communal areas by veterinary cordon fences (Scoones and Wolmer 2006).

Because Botswana has successfully gained area-based disease freedom from communal areas and provided regional and export marketing opportunities to marginalised pastoralists many believe this approach can, and should, be replicated in areas of Africa where endemic diseases currently limit trade and export (cf. Kivaria 2003). Officials from livestock sections across sub-Saharan Africa are increasingly looking at Botswana as a disease management and marketing success story, including disease-free zone formation into ongoing rural development planning. Whilst there are signs that this approach may become increasingly less attractive for several economic and political reasons (see, e.g. Scoones and Wolmer 2006; Scoones et al. 2010), the nation's fences provide a unique opportunity to investigate what impact this policy might have upon the pastoral production system.

Within Botswana, Ngamiland district in the northwest of the country provides an excellent opportunity to evaluate the impact of fences constructed in living memory because it is often at the 'frontline' of government disease control efforts and a priority region in terms of disease control. The district is one of the most remote and inaccessible district in Botswana and its close proximity to areas where diseases such as cattle lung disease (or Contagious Bovine Pleuropneumonia, hereafter CBPP) and foot-and-mouth disease (hereafter FMD) remain endemic in rural livestock and wildlife make the region a target area for investment in veterinary cordon fencing. Nine of the country's 18 documented disease outbreaks have occurred in the region (Hitchcock 2002), and, excluding fences put up over the last few years, the last five of the country's 17 major veterinary fences where constructed in the district (Figure 1).

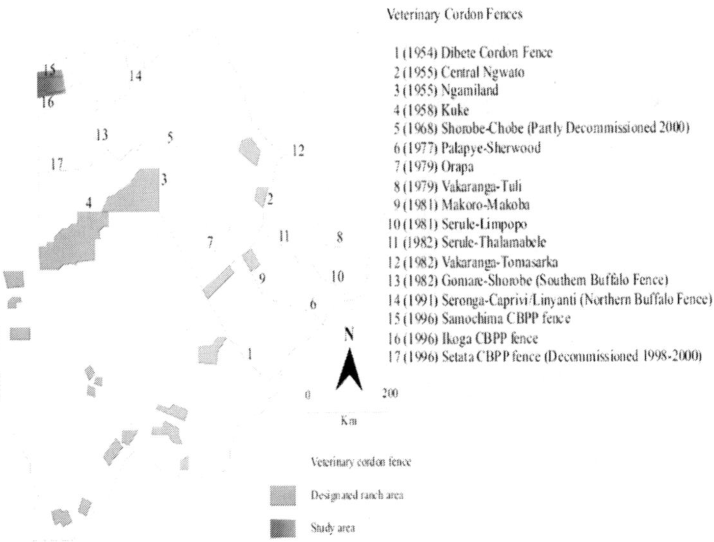

Figure 1: Botswana's veterinary cordon fences and fenced ranch blocks on 1954 to 1996. The area marked in red represents the field site used in this project (Source: Adapted from Thomas and Shaw, 1991).

Chukumuchu Village

Chukumuchu is a small settlement located within a sparsely populated area between the Okavango River and fenced international border with Namibia (Figure 2). The village itself is home to approximately 270 people, whilst a further 230 or more herders reside at cattleposts dispersed across the surrounding rangeland (Republic of Botswana 2001). The area remained largely unfenced until an outbreak of CBPP in 1996 when several veterinary fences were rapidly erected in an effort to contain the disease. Two of these, the Ikoga and Samochima fences, now enclose the settlement to the north and south. Livestock movement is only permitted north through the fences via a few manned gates. To do so, herders must first apply for a permit, which can be obtained from the village when veterinary officials are present (usually during biannual vaccination campaigns) or from the district veterinary office at the nearest town; a return trip of over 100 km. From a disease management perspective the area falls within the 'red zone' where cattle are vaccinated for FMD twice a year and cannot enter the export chain.

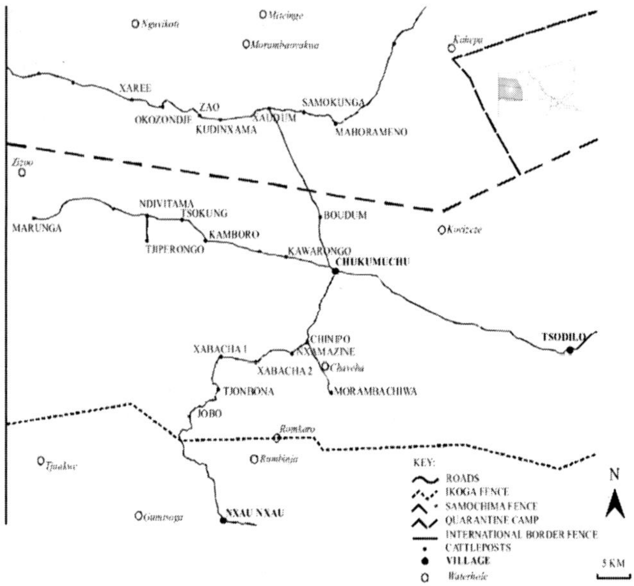

Figure 2: Chukumuchu village and surrounding waterholes and pans available to pastoralists in the past. Source: Fieldwork interviews (note: the location of all waterholes except Romkaro are approximate, based on map-supported interviews).

The climate is semi-arid with an average annual rainfall of between 400 and 500 mm. Over the last 25 years, annual rainfall has been as low as 260 mm and as high as 780 mm. The landscape is typical of the Kalahari with the undulating topography dominated by the presence of Kalahari-age vegetated longitudinal dunes orientated in an east-west direction, interrupted by several fossil river valleys, clay pans and depressions. The shrub and tree savannah varies from Burkea african, Pterocarpus angolensis and Baikiaea plurijuga communities typically found on sandy dune ridge soils to patches of interdune valley woodland dominated by Acacia spp. and Combretum imberbe.

The main ethnic groups are the pastoral Herero and the agropastoral Hambukushu. For the former group, Lake Ngami to the south represented the easternmost point of their territory until their numbers increased dramatically following the German-Herero war from 1904 to 1907 (Werner 2000). From the 1930s onwards, the group have spread northwards into the area, having rebuilt their herds (Pennington 1992). According to Herero and Hambukushu herders, they expanded pastoral production into the region from the 1950s, digging a series of dry season wells along low lying valleys.

The Research Methods

Pastoral systems are dynamic, necessitating a multitemporal approach to the analysis and understanding of change (Desta and Coppock 2004). Whilst some characteristics, strategies or cultures may remain relatively consistent, most communities are constantly adapting and changing to a variety of internal or external pressures. History, particularly historical struggles over access to, control of and exclusion from resources are vital in the quest to understand current resource use and management issues (Moore 1993). To investigate the impact of veterinary cordon fences, this study sought to understand how mobility was used in the past, how it is used now, along with the timing and cause of its demise. Only within this narrative of social change can the direct impact of veterinary cordon fence enclosure on pastoral livelihoods be illuminated.

Research focused on documenting social changes relating to resource access and mobility within pastoral society over the period leading up to, directly following, and the years since the area was enclosed by fences in 1996. It was considered particularly important to understand how such structures influence aspects of livestock management in a contemporary context, given both the internal dynamics described above and likelihood of future expansions to the fence network. The study was therefore intentionally situated in a region where fences have been constructed 'in living memory' so as to explore their impact upon contemporary livelihoods and natural resource use. Fieldwork was conducted from 2005 to 2006 and combined a variety of qualitative and quantitative methods within a multi-method approach (cf. Madsen and Adriansen 2004). Oral testimonies provided the main source of data, recorded during a series of focus group discussions, in-depth semi-structured interviews or informal discussions. Households and individuals interviewed at length were selected on the basis of a short community questionnaire survey (n = 53). This recorded data on basic socio-demographics, present livelihood activities, settlement history, and past and present livestock mobility. Further questions were also posed regarding local opinions of the fences, understandings regarding their purpose, along with local views regarding the work of the veterinary department. Emerging information concerning contemporary herd management and the use of natural resources in the past was complemented by direct observations during a period spent participating in day-to-day cattlepost management with key informants (i.e. herding, milking and watering livestock). Where the location of key resource areas could not be confirmed in the field due to logistical difficulties (i.e. remote inaccessible locations), the approximate position was estimated using natural resource mapping exercises aided by a printed Landsat TM satellite image (Theis and Grady 1991). This combined approach allowed the issue of natural resource enclosure to remain at the centre of analysis, whilst thoroughly investigating the local-level context of natural resource access and use. To protect the identity of herders on what has historically been a contentious issue, I use initials to refer to key informants throughout this paper.

Pastoralism and Mobility in Northwest Botswana

Existing Literature on Herero Mobility

Like most pastoral groups, Herero society is known to have changed dramatically over the past century. Prior to their forced migration to Botswana at the beginning of the twentieth century, the Herero were described by early missionaries as 'inveterate nomads' living a 'constantly nomadic life' moving from 'place to place with their herds, depending on the conditions of the pastures' (Werner 2000: 249). Anthropologists researching Herero in northern Botswana during the early 1980s compared their cooperative cattle management systems and transhumant land use practices to mobile pastoral groups from East Africa (Smith 1992). Other authors, however, suggest that Herero settlement patterns were more permanent than neighbouring Himba pastoralists and in comparison to the temporary camps and mobile arrangements of East African Turkana or West African Fulbe, the group's social organisation is much more tied to one place (Bollig and Gewald

2000).

Bollig and Gewald (2000) also note that the Herero use far less intensive herd management practices than East African pastoralists, usually herding livestock in the direction of intended pastures in the morning and relying on cattle returning at dusk for water. This form of 'herd-release' management is in fact the traditional cattlepost system employed throughout the Kalahari by Herero, BaKalagadi, Tswana and Mbukushu herders alike; to avoid working through the extreme heat of the day and generally based on the minimum expenditure of energy (Perkins et al. 2002). What differentiates the Herero from other Kalahari pastoral groups like the Tswana is their social organisation, which is clearly linked to their systems of opportunistic management and mobility. Central in this regard is their main homestead or onganda where several patrilineal-related households reside together. Each social group also own another homestead or temporary cattle camp at a waterhole where some of the men and their wives move with the cooperatively managed herd during the wet season (Almagor 1980).

The Herero could, therefore, be described as semi-sedentary pastoralists utilising opportunistic management and mobility in order to cope with, and take advantage of the highly variable resource base (Almagor 1980). Whilst the Herero may have been more mobile in the past, the main onganda has remained sedentary since at least the late 1970s in Botswana (Vivelo 1977) and for over 58 years in Namibia (Bollig and Gewald 2000). However, the Herero may still use mobility as a principal drought-coping strategy and herd management technique. Since the late 1970s in Namibia, Herero pastoralists have been known to truck or trek cattle from one grazing reserve to another during prolonged droughts (ibid: 28). In Botswana, the Herero objected virulently to enclosure by the proposed tribal grazing lands policy (TGLP) ranches near Lake Ngami, largely on the grounds of the reduced mobility this would bring (Almagor 1980).

Present-Day Livestock Management Surrounding Chukumuchu

Though the Herero may have retained some form of pastoral mobility, there are further social and economic changes that have occurred within Herero society since their (re)pastoralization and expansion in northwestern Botswana. Today, livestock production in the region is centred on a series of permanent hand-dug wells situated along interdune valleys. At almost every cattlepost, at least one owner possesses a 2000-l plastic water tank, petrol or diesel generator and electric water pump. All of this equipment can be purchased at hardware stores in the nearby town. For those who lack the resources to purchase this equipment, water can be accessed either for free at a government borehole near the village; through formal payment to a principal well owner; through informal herding arrangements where access forms part of payment in kind; and finally at private wells operated by hand, using buckets lowered and raised on a winch system (for those fortunate enough to be gain a permit from the land board and wealthy enough to afford the construction costs).

A second significant change from the pastoral system described during the 1980s is the region's social organisation. With the mechanisation of most water sources very little labour is now required at the cattleposts. Most households have their main homestead in the village where children attend school and family members often gain informal income from a variety of sources (e.g. brickmaking, thatching, dress making, labour-intensive public works programmes, beer sales). Typically, one or two males remain at the cattlepost to milk, water and kraal the livestock daily. A few wealthy farmers own a four-wheel-drive vehicle which is principally used to collect fuel from the nearest town to run the water pumps. Several cattleposts were managed by contract herdsmen for absentee owners living in the nearby town of Shakawe. At one cattlepost, a single herder was found to be managing several herds for different absentee owners.

Pastoral Mobility and Social Organisation 1950-1980

Because the present nature of pastoral mobility and social organisation in northwestern Botswana clearly diverged from that documented by anthropologists in the 1970s and 1980s, significant research effort was directed towards documenting changes to pastoral mobility and management from the period pastoralists moved into the region, to the present day. According to (Pennington 1992), Herero pastoralists have been expanding their territory to the north of Lake Ngami since the 1930s predominantly along the western margins of the Okavango

Delta. Their numbers remained low in the dry Kalahari sandveld to the west of the Delta fringe, however, until the 1950s after which Herero settlement increased (Lee 1984).

Oral testimonies from several elderly herd patriarchs interviewed in the Chukumuchu area confirmed that Herero and Mbukushu households moved into the region from the 1950s onwards. Most cited stock losses to predators along the Okavango River and shortages of grazing are reasons for their migration inland and they clearly described how they expanded first along the Xaudum valley where water could be easily accessed from shallow pit wells. By the 1960s, there were wells throughout the valley and cattleposts had begun to be established at large waterholes in the interdune valleys between Xaudum and Nxau Nxau. One of these was established at Chukumuchu, a large waterhole retaining water long after the wet season. Chukumuchu remained a large Herero cattlepost until in 1991 when a government settlement was formed at the site.

Discussions about past herd mobility and management revealed that herders could be divided into three groups according to the type of seasonal livestock movement practised from when the area was first occupied by pastoralists in the 1950s to the early 1980s. First, some people with small herds of cattle, mostly of Mbukushu origin, have never practised seasonal mobility and preferred to fence fields against livestock damage and practise 'herd-release' or 'free-range' management. Second, pastoralists engaged in arable agriculture with larger herds practised regular seasonal moves to a particular waterhole during the growing season partly to relieve pressure on grazing and water, and partly to avoid crop damage. Finally, pastoralists with large herds, primarily of Herero origin, would make seasonal camps at waterholes located within a certain pasture area, sometimes utilising more than one waterhole per wet season and sometimes visiting different pasture areas in different years.

There are, therefore, two groups of pastoralists who conducted seasonal herd movements in the past. For those with arable fields, mobility was primarily conducted to protect arable lands from livestock damage and reduce the need to rigorously maintain labour-intensive thorn fencing. 'We moved the cattle to our camp at Chaveha just after the rains to stop the cattle destroying the crops (November 2005)', said MT an elderly Hambukushu man from Chukumuchu. According to MT, he would herd the cattle with his sons and make a temporary camp at the waterhole each year, remaining until the water dried up, sometimes as late as June or July. Meanwhile, the women and remaining children would look after the fields (bird scaring, harvesting, threshing) and attend to small stock. A similar pattern was practised by KM, a Mbukushu man from the cattlepost Xabacha II. KM used to camp at the Romkaro waterhole, now dissected by the Ikoga fence, which he says held water for up to 3 months after the rains had stopped.[b]

Footnote

[b]Interview with KM (November 2005)

The second form of seasonal mobility practised in the past was that of large herd-owning Herero pastoralists. For Herero herd patriarchs, the primary motivation was to alleviate the resource pressure on grazing surrounding permanent cattleposts. Another reason to move was the ease of watering livestock at waterholes as this requires far less labour than permanent hand-operated water sources. In an annual cycle, Herero herd patriarchs would trek cattle to temporary camps near one or more large waterhole, from which livestock would be managed using the usual 'herd-release' method. Pastoralists would generally move with their livestock after the first major rainfall event of the wet season (usually between October and December), remaining until the water sources dried out (usually May or June). Typically, male household members or contract herders would move to cattle camps, whilst wives and extended family would remain behind at the main cattlepost to look after smallstock and a small milking herd. Some Herero women would visit the camps to make butter and oil from milk, as they report that yields were good during the wet season. Most households would therefore utilise waterholes no further than a day's walk of the main dry season homestead. These arrangements are similar to those noted by (Almagor 1981) for Lake Ngami Herero.

Although herders could provide the name and approximate location of at least four or five waterholes regularly used within 5-10 km of their dry season cattlepost, most would spend the entire wet season camped at a single waterhole. For example, KK, a Herero herd patriarch and owner of Jobo cattlepost, described moving his herd over different years to Ramkaro, Rumbinja, Gunisoga and Tjaakwee waterholes (Figure 2). All of these are now partly or fully enclosed by the Ikoga fence. Herders such as KK would vary the waterhole visited mainly on the basis of where the rains were heaviest[c] This is a similar pattern of mobility to the 'meeting the rain' management described by Adriansen and Nielsen (2002) for Fulani pastoralists in West Africa.

Footnote

[c]Interview with KK (October 2005)

Although the precise years when mobility declined were difficult to establish, most herders cited the early 1980s as the time when they became fully sedentary. For pastoralists with large herds, the transition to sedentary management often involved constructing a second permanent well at their favoured wet season waterhole. This allowed greater flexibility enabling the owner to split the family herd, use mobility in emergencies or continue to make regular seasonal moves to one specific area. This restructuring of pastoralism is considered in relation to the fences in more detail below.

Once the nature of past mobility was established, the questioning centred on the main factors driving sedentarisation. These enquiries revealed several main forces causing sedentarisation. First, an out-migration of households and farm labour to regional towns and large villages caused by regional socio-economic developments. Second, a gradual erosion of the traditional Herero household structure and cooperative herd management system. And finally, a period of Herero repatriation to Namibia from 1993 to 1996, in an attempt by some to regain Herero political autonomy. All of these changes caused a reduction in the availability of labour at the region's cattleposts and ultimately reduced herd mobility. In the following sections, each of these forces are briefly described in turn, before examining the situation regarding pastoral mobility and opportunistic management which existed at the time of the CBPP outbreak, and the subsequent impact of the fences on pastoral management.

Pastoral Sedentarisation Pre-1996

Out-Migration and Settlement in Villages or Towns c. 1970s-1980s

One of the main factors causing pastoral mobility to decline during the 1970s and 1980s was labour shortages at the cattleposts. In response to the question, 'Why do you no longer move your cattle seasonally?' one Herero herder replied simply, 'I can't take the cattle away to waterholes as I don't have enough herd boys to help and my wife is usually in the village so there would be no one to look after the smallstock' (November 2005). The reduction in labour availability has been driven by an ongoing process involving the settlement of household members in nearby villages or towns. This followed improvements in the quality of public services and various physical capital assets in the region, as roads and villages have developed through government investment in rural development. The expansion was accompanied by improved social welfare and livelihood opportunities in larger villages through increases in waged labour, education, health care and trade. In Ngamiland district as a whole, these trends can be seen in the disproportionate increase in the population of the district's main town Maun, whose population has increased more than three times the rate of the district as a whole (Bensen and Meyer 2002). In the Chukumuchu region, household members started to reside in large villages such as Shakawe and Nxamasere from the late 1970s to 1991 when the village of Chukumuchu was established by the Botswana government. Since this time Chukumuchu has steadily expanded with the construction of a school and clinic which attracted further out-migrations of rural households and resulted in the bilateral social organisation observed in the study area today.

For Herero pastoralists, the socio-economic changes documented above represented an erosion of their social organisation, which had major implications for herd management and mobility. According to local herd patriarchs, elements of the patrilineal homestead or onganda remained intact in the region until the early 1980s, which is in agreement with anthropological research of the time (see, e.g. Almagor 1981). However, the Herero household structure had already started to become increasingly compromised as more generations of the Herero sought outside employment and education. As a result, many Herero families began to establish their main homesteads in regional towns and large villages and started to manage their cattle from a distance, leaving mainly older males and unemployed young men at the cattlepost. KE, a Herero well owner from the Xaudum valley, described the changes that occurred within Herero society:

'During the 1970s many of the Herero family members became educated and found work in the towns. By the late 1970s most of the mothers and children had left the cattleposts and moved to Shakawe or Nxamasere and in the early 1980s many families stopped making seasonal movements to waterholes (November 2005)'.

Most households stopped going on seasonal livestock movements due to the lack of extended family members able to remain at the main cattlepost and tend to calves and smallstock. Whilst cattle can often be left to roam freely in an area they know, small ruminants require daily herding because they can easily get lost, stolen or taken by predators (Adriansen and Nielsen 2002). In most pastoral societies, cattle are usually the responsibility of the household head, whilst small ruminant herding is the responsibility of children or women. The change in Herero household structure caused many to reduce the number of smallstock kept at the cattlepost as they could no longer be looked after by children and female members of the household. Once again, this trend is supported by anthropological observations amongst Ngamiland Herero in the late 1970s (Almagor 1981). During the 1970 s and 80 s, therefore, pastoralism for the Herero in the Chukumuchu rangelands was characterised by declines in labour availability and reduced pastoral mobility. However, a further decline in the number of pastoralists likely to practise seasonal mobility occurred after Namibian independence in 1990.

Herero Repatriation 1990-1996

The Herero of Botswana have often been described as a proudly independent ethnic group with aspirations to return to their homeland in Namibia (Pennington and Harpending 1991, Almagor 1981). These aspirations came into fruition following Namibian independence when the governments of Botswana and Namibia agreed on an organised repatriation of Herero pastoralists to Gam in Namibia (Kruger 1998). Whilst official data regarding the scale of the Herero repatriation from the Chukumuchu region are lacking, the decline in Herero households was evident from the number of newly established Hambukushu and Bayei herders in the region and several abandoned and vacant Herero wells.

Almagor (1981: 49) noted that the Herero were closely monitoring political developments in Namibia as early as the late 1970s and were reluctant to invest in borehole technology or deepen their economic involvement at the time as a result of 'encouraging signs of an approaching independence.' Today, most Herero engaged in livestock production in the study area have invested in water pumping technology or village-based homes and many have become more involved in the local village institutions and regional politics. SM, a Herero man in his fifties, for example, is the local councillor for the region and a prominent figure in the community. According to SM before the cull he owned over 100 cows and used to practise seasonal mobility. After the cull, he decided to keep fewer cattle and invest in water pumping and storage equipment,

A large herd need a lot of water, its hard work watering by hand, so you must move to search for waterholes in the wet season. I wanted to stop having to move around so much so I invested my compensation in a water tank and pump, you can't move these things easily (October 2005).'

It would appear, therefore, that the Herero repatriation resulted in the loss of households less integrated into regional society, less likely to have invested in borehole technology or sedentary infrastructure, and more likely to use mobility. For these households, mobility was an important part of management as they preferred to 'continue to rely on the natural resources' and not invest in 'expense and activity which would deepen their economic involvement in a country which they see as their temporary host' (ibid: 49).

Social changes in the region from the early 1980s to the early 1990s therefore caused some fundamental alterations to pastoral management and herd mobility. Pastoralists faced with an out-migration of household members to large villages and labour shortages were forced to decrease their herd mobility. The Herero repatriation may have also resulted in the loss of many of the most mobile pastoralists. Although some herders may have begun to adapt to these changes before the outbreak of CBPP in 1995, the CBPP cull caused an enormous disruption to the pastoral economy which resulted in a major restructuring of pastoralism in the Chukumuchu region.

The CBPP Outbreak 1995-1996

In February 1995 Botswana's first CBPP outbreak for nearly 60 years occurred in the Xaudum valley to the north of Chukumuchu.

The outbreak started after infected livestock entered Botswana from Namibia, probably as a result of the post-independence relaxation of military border patrols and the ongoing process of Herero repatriation. State veterinarians moved quickly to contain the disease, using emergency powers to construct three new veterinary cordon fences. These failed to hold the outbreak and in April 1996 the Department of Animal Health and Production decided to 'stamp out' the infection by ordering the cull of Ngamiland's 320,000 heads of cattle.

The depopulation of livestock in the Chukumuchu region, and throughout Ngamiland, caused the migration of many households and the majority of rural farm labourers to village service centres or regional towns. This migration, and the provision of compensation and relief work which followed, forced the rapid acceleration and consolidation of many longer-term tensions which had hitherto been driving changes within the pastoral management systems of the Chukumuchu region. Firstly, many households took the opportunity to establish a second homestead in the village, especially as there were widespread rumours that the Chukumuchu veterinary zone would be restocked first. For example, 12 out of 24 households questioned in the village migrated to Chukumuchu and established homesteads in 1995 and 1996. Many of these households migrated from the infected red zone to the north of the Samochima fence, with seven of these 12 migrating from the Xaudum valley.

The CBPP cull and subsequent migration of households to large villages caused enlightenment among young herdsmen who quickly realised the benefits that commoditisation had brought to service centres and soon became entrenched in village life. According to one young man,

'After the cull there were so many young people just sitting around in big groups enjoying themselves. The herd boys realised what they were missing and many decided then not to go back (October 2005)'.

The CBPP cull, therefore, prompted a further out-migration of potential herders who soon became engaged in alternative income opportunities within large villages and towns. The decline in pastoral labour has been widespread in Ngamiland. This can be seen in the official statistics for formal employment in the region. In 1991 the agricultural sector was the largest employer with 40% of the district's labour force, whereas only 14% were employed in 2001 (Bensen and

Meyer 2002). The majority of households became reliant on emergency food rations and labour-intensive public works programmes for income (ibid, p.17).

For pastoralists, the CBPP cull and subsequent compensation payments provided a huge increase in the availability of cash for the household economy. After the cull, the government offered several compensation options, from all cash compensation (compulsory for small farmers with up to ten heads of cattle) to 70% cash and 30% restocking, or 30% cash and 70% restocking. Regionally, most farmers opted to receive high rates of cash compensation (Fidzani et al. 1999) [d] In the Chukumuchu region 43.4% of households surveyed opted for 100% cash compensation and the remaining households opted for compensation involving either 30% or 70% restocking and the rest cash[e] Most households spent the money on immediate or routine household expenditure, especially households with small herds of cattle. However, several households choose to invest their money on major purchases such as vehicles and water pumping technology. These purchases were a direct adaptation to the pressures forcing sedentarisation (i.e. labour shortages) and represented the beginning of a period of agricultural restructuring in the Chukumuchu rangelands.

Footnote

[d]Fifty-four percent (54%) choose 100% cash compensation and 23% choose 70% cash compensation (Fidzani et al. 1999).

[e]Source: fieldwork survey (n = 53)

Ngamiland's pastoral lands were restocked between April 1997 and March 1998 with 70,000 cattle (Mullins et al. 2000). As discussed earlier, once they received their livestock, pastoralists in the Chukumuchu veterinary zone could only move livestock up through the zones with the requisite permits from the Department of Animal Health and Production (DAHP). As a result, many abandoned shallow wells previously used in the Xaudum valley to the north, in favour of wells located along the interdune and dry river valleys of the Chukumuchu region. A period of pastoral restructuring ensued as households reorganised their livestock management arrangements within the new veterinary disease control zone. Pastoralists with the financial resources to employ a contract herder now had to entice labour away from increasing employment opportunities available near villages and urban centres. Whilst households with cattleposts within close proximity to the village could rely on young family members to provide herding labour after school and at weekends, owners of remote cattleposts struggle to find contract herders willing to work in the rangelands. Several herd patriarchs spoke of matching the minimum labour-intensive public works wage and providing extra incentives such as food and other daily consumables, in order to attract labour. The increased investment in water pumping technology was also designed to make the herding work easier and more attractive to contract herders.

Herder Opinions of the Veterinary Cordon Fences

Given the trends towards sedentarisation and management restructuring that occurred amongst the region's pastoralists in the years leading up to, and directly following enclosure in 1996, it comes as no surprise that on the whole, few pastoralists were opposed to the fences. A survey of 53 households from both the village and surrounding cattleposts posed questions regarding the veterinary work conducted by the DAHP; local understanding of the purpose of the veterinary fences; and local perceptions of the positive or negative impacts of veterinary fence enclosure. For the latter question, the survey group was asked to provide their overall opinion of enclosure (i.e. positive or negative), as well as the main reasons for their stance on the fences.

Based on the data collected from the village of Chukumuchu (as opposed to the surrounding cattleposts), the veterinary work and services provided by the DAHP are well received and perceived as positive to local livelihoods. Ninety-six percent of informants in the village consider the work of the DAHP as providing positive benefits to livelihoods, enjoying the free vaccinations and veterinary disease control (Table 1). Complaints over the frequency of the vaccinations and lack of community members employed by the DAHP concern the 4% who see the veterinary work as negative, with informants suggesting that vaccinations could be done less often to avoid disruption to local lives.

Table 1: Contrasting opinions of DAHP work and CBPP fences from village and cattlepost-based households

Region of survey	Opinion of DAHP work + CBPP fences	Positive (%)	Negative (%)	Neutral
Chukumuchu (*n* = 24)	DAHP	96	4	0
Cattleposts (*n* = 29)	DAHP	79	14	7
Chukumuchu (*n* = 24)	CBPP fences	83	13	4
Cattleposts (*n* = 29)	CBPP fences	21	59	20

Source: survey of 53 households in the Chukumuchu regionMcGahey Pastoralism: Research, Policy and Practice 2011 1:14, doi:10.1186/2041-7136-1-14

With regard to the CBPP fences, 83% of the villagers think that the fences provide positive benefits to livelihoods, when given the option between positive, negative or no change. From the remaining pastoralists, 13% see the fences as having a negative impact on livelihoods, often stating that the fences reduce livestock grazing, prevent the movement of meat, and cut off shortcuts to regional villages and cattleposts. When the 83% of informants who see the fences as positive were questioned about their reasoning, 54% replied that the fences stop disease spreading from the north and 42% stated that the fences stop cattle from straying and reduce stock theft (4% saw no positive impact).

It would, therefore, appear that the fences are generally well received in the region and only cause a minor inconvenience to local livelihoods. However, the survey data from herders at cattleposts in the region surrounding Chukumuchu reveal highly contrasting results. Whilst there are similar opinions regarding the work of the DAHP (79% see the veterinary work as beneficial), there is a mixture of opinion regarding the impact of veterinary fences as 59% of informants regard the fences as negative. Furthermore, whilst 21% of informants from the cattleposts see the fences as positive, 20% see no positive or negative impacts. Those who see the fences as negative cite blocked short cuts (7%), restrictions on the movement of meat (7%), and reduced livestock mobility (38%) as reasons. The remaining complaints are largely against the lack of maintenance following wet season elephant damage (7%), rather than conflicts over resources and movement.

The results of this survey illustrate that there is a clear divide between the views of those living within Chukumuchu and those residing at cattleposts within the surrounding rangelands. Given that maintaining access to cattlepost resources is of greatest interest to middle and better-off herders, and that these pastoralists were often looking after the interests of their village-based family, this contrast was significant and represented the views of those most likely to be experiencing the opportunities and constraints of enclosure. Further analysis of views revealed that concerns can be divided between the 14% of cattlepost dwellers who complain about what, initially, appeared to be minor difficulties caused by enclosure and those who cite longer-term issues such as reduced mobility.

Most people in the region have ties to land or resources found to the north or south of the CBPP fences and enclosure has caused difficulties for people attempting to conduct their usual livelihood activities. Take, for example, the case of DS. DS is 65-year-old Mbukushu woman from Chukumuchu and her husband manages their family herd at the Morambachiwa cattlepost 9 km to the south of the village. DS also owns cattle at another well in the Xaudum valley to the north of the Samochima fence. DS complains that the fence has made it harder for her to herd cattle between Xaudum and Morambachiwa which she occasionally needs to do, especially after water or grazing become scarce.

'Before the fence I could go alone and just herd the cattle back, now I need a permit and have to employ a herd boy as we have to stop at the fence gate and fill out forms (October 2005)'.

She also complained about the problem of moving meat across the fence, 'if one of my cows die at Xaudum I can no longer bring the meat to my house in Chukumuchu (October 2005)', she continued.

Whilst those who retained access to wells and grazing to the north resent not being able to take meat over the fence or freely move livestock, there are other people who have links to the village of Nxau Nxau to the south of the Ikoga fence. A 58-year-old Herero man from Jobo cattlepost called KK, for example, explained his problems attempting to provide milk and meat for his wife and daughters who live in Nxau Nxau:

'Before the cull I planned to dig a well close to Nxau Nxau so that I could split off some cows and a few goats for my wife and daughters so that they can have some milk and meat. The land board granted a permit before the cull but now they have taken the permit away as they don't allow wells within 3 km of the fence. Now if I want to take a goat to slaughter at Nxau Nxau the vets have to come and inspect the stock and it takes a long time (October 2005)'.

Some people complained about the Ikoga fence blocking direct donkey tracks used to travel between cattleposts and Nxau Nxau village. Most of these complaints were from people at the Xabacha or Nxamazine cattleposts and concerned the added distance required to travel through the Jobo veterinary gate. One old man from Nxamazine explained, 'it's as if my direction is now towards Namibia (October 2005)' due to the need to travel west before heading south to the village. KM a 35-year-old Mbukushu man from the Xabacha II cattlepost explained what happened after the cull when he found that the Ikoga fence now blocked his track:

'Before 1995 we used to go straight to Nxau Nxau on donkey and it would take us from sunrise to about 1 pm to reach the village. After they killed our cattle we went one day on our old track and found the fence blocking our track. We were very angry and it took all day to get to the village. Why didn't they put a small gate for us to use? They must have noticed our track? (October 2005)'

When pressed to provide a negative implication of enclosure most pastoralists could offer one of the above problems. These negative impacts are similar to those documented by Hitchcock (1995,2002), who noted that movement restrictions across the central Ngwato cordon fence during a FMD outbreak prevented the seasonal movement of cattle to village arable lands causing a shortage of milk and draught power. Hitchcock (1995) also described how officials prevented people taking meat and crafts to markets found to the east of the fence by restricting the movement of some goods of animal origin (ibid, p. 184).

During interviews with individual herders whose overall view of the fences is positive or negative, some longer-term concerns emerged, however. Furthermore, pastoralists in the region could be divided into three distinct groups based on their positive or negative position on the fences. The majority of herders concerned about reduced mobility, for example, were either Herero herd patriarchs >50 years or young herding households with male household heads <40 years. Conversely, of the 21% who claimed that the fences were positive, the majority welcomed the added enclosure on the grounds that the fences reduced stock loss through cattle straying, and these pastoralists were mainly recently established Hambukushu or Bayei herders. In the discussion below, the views of three households representing a composite of each of these groups are presented.

Welcoming Enclosure: Recently Established Mbukushu Herders

Most pastoralists who viewed CBPP fence enclosure as positive were recently settled Mbukushu herders typically aged between 40 and 50 years old. In many cases, these were households that had used the Herero repatriation as an opportunity to gain access to water and grazing. The case of KB, the Hambukushu owner of Tjiperonga cattlepost, was representative of this group. KB purchased the well from a repatriating Herero herder in the early 1990s and spends most of his time at the cattlepost with his eldest son whilst his wife and extended family live at a second homestead in Chukumuchu. Like many pastoralists in the region, KB owns a petrol generator and electric water pump, along with a plastic water tank. Whenever he or his son are alone at the cattlepost they use the generator and pump to provide water for the livestock, as the bucket and winch system requires at least two people to operate productively.

Labour shortages at the cattlepost concerned these herders and is the primary factor leading to their positive position on the fences as KB explained:

'As you can see I cannot afford to pay a herd boy and rely on my eldest son to help me at the cattlepost while my other children attend school in the village. The fence is good and it stops our cattle straying very far, I can now easily find them as I know they can only go as far as the fence (October 2005)'.

In the dry Kalahari, livestock are physiologically constrained to graze within several kilometres of drinking water, especially in the dry season (Moleele and Perkins 1998). Under the traditional cattlepost system, the borehole or well has been described as effectively 'the herder' due to the reduced need for herding during the dry season (Jerve 1982). Like much of the Kalahari, in the Chukumuchu region there are many ephemeral waterholes or pans often situated along interdune or fossil river valleys. As a result livestock often walk considerably large distances in search of fresh pasture in the wet season as they access these seasonal water sources. Perkins (1996) notes that straying, especially after the first rains, results in much time and effort spent by herders retrieving cattle from distances of up to 40 km away. Many pastoralists are, therefore, aware that the new CBPP fences and veterinary zones now present a barrier to livestock who can now only stray as far as the Samochima fence to the north, the Ikoga fence to the south, or the Namibian border fence to the east.

Thus, the CBPP fences have benefited herders with limited human capital and this is reflected in the 21% of herders from the cattleposts and 83% of herders from the village who recognise the positive outcomes of enclosure[f] Veterinary cordon fences and their associated maintenance roads, sporadic gate camps, along with regular maintenance patrols and permanent gate keepers also represent infrastructure and personnel that can be utilised by pastoralists. When cattle fail to return home after release in the wet season, pastoralists begin looking for livestock often relying on the knowledge of other herders who may have seen their cattle in the rangeland. Cattle often stray as far as the veterinary fences where they can be easily seen or tracked. Pastoralists use fence roads to search for livestock, often stopping to ask DAHP gatekeepers if their cattle have been seen along the fence. DAHP officials may also visit cattleposts to inform herders about the location of lost herds.

Footnote

[f]Human capital refers to the labour available to the household and changes constantly due to internal demographic reasons (e.g. births, deaths etc.) and to deliberate restructuring to meet unexpected events (e.g. divorce) or external pressure (Ellis 2000).

Beyond reducing stock losses through straying, some pastoralists report that enclosure by the cordon fences has improved livestock security in the region. MS a male Hambukushu herder in a similar situation to KB explained why he welcomes enclosure:

'The fences are good, especially the Samochima fence to the north. Before the fences were made there used to be a lot of cattle stolen to the north and now it has decreased. In the past people would steal your cattle during the night and take them to Namibia.(October 2006)'

In Namibia, large tracts of communal rangeland have been compartmentalised and fenced illegally by individuals and small groups of farmers (cf. Werner 2000, Stahl 2000, Stahl 2000) found that reduced stock theft was one of main benefits of illegal fencing and indeed the main determinant driving the impetus to fence. Before the CBPP fences and additional electrified international border fence were constructed in 1995/1996, there was a period after Namibian independence in 1990 when military border security was relaxed and cattle thefts increased in the Xaudum area, probably precipitating the CBPP outbreak itself. It seems for many the cordon fences provide a welcome barrier preventing cattle from straying into remote border regions where the risk of theft remains high.

Increasing Vulnerability: Herero Herd Patriarchs

Whilst many recently settled Hambukushu and Bayei herders may welcome enclosure, a small group of Herero herd patriarchs remain openly negative towards the cordon fences. This group cites reduced livestock mobility as their primary concern. The views of KB a 60-year-old Herero herd patriarch who manages the household herd at the Boudum cattlepost are representative of those expressed by this group. KB and his family were one of the first herders to pioneer livestock production in the region and he remembers moving to the Xaudum valley to dig a new well with his father in the late 1950s after losing cattle to trypanosomiasis and predators at Nxamasere. In 1984, the family decided to move their permanent onganda to their wet season waterhole, Boudum, as grazing had been depleted at Xaudum due to the ongoing drought. The family kept their cattle permanently at Boudum until 1995 when poor rains and lack of grazing forced KB to move the herd to his brother's well at Xaudum where they remained until the cull.

Since their cattle were restocked in the Chukumuchu region, like many households, they have remained in the area since 1997/1998 and established a village-based homestead. When asked his views about the CBPP fences KB replied, 'it seems as if I'm in a ranch, I can't breath, I can't move anymore'. During discussions with KB and others like him, it was clear that this group were more concerned about the lack of mobility in times of emergency rather than the enclosure of seasonal resources. KB explained why he resented the CBPP fences,

'Before the fences were made you could take your cattle anywhere you wanted. If there was good rainfall at Xaudum you would go there. Before if my well dried up or there was a drought I would be able to go to Nxau Nxau or Xaudum and ask a friend or relative to keep them there. Now, since the fences were made if my well dries up I will have to stay here and my cattle will die here (November 2005)'.

KB's response to enclosure highlights some of the contemporary risks and vulnerabilities facing sedentary pastoralists in the region. The strategy he used to cope with risk in the 1995 drought also illustrates the contemporary context of mobility used before enclosure to mitigate these problems. According to KB, during droughts, his well often runs dry and he is forced to try and dig deeper. Before enclosure by the cordon fence, he could draw upon social ties to access water sources elsewhere in the region. Today, with movement restrictions and the permit system, livestock cannot be easily moved across the fences[g].

Footnote

[g]Cattle can be moved in a northerly direction across the fences but only after applying for a movement permit. This can take some time as veterinary officials only occupy their village-based office during vaccination campaigns and pastoralists must often travel to Shakawe and visit the district office. Few animals have been moved south over the fences as this would require either a veterinarian to visit the cattlepost and inspect stock or livestock to spend 21 days in quarantine which would incur fees.

These mobility issues are similar to those faced by sedentary pastoralists enclosed within TGLP ranch areas. Perkins (1996) notes that breakdowns are a frequent occurrence with mechanised boreholes in the Kalahari. Borehole owners often have a network of boreholes and/or contacts and use herd movements, often over large distances, to alleviate breakdowns, droughts and spatial variations in grazing (ibid: 505). Although many herders in the Chukumuchu area owned water pumping and storage equipment, most wells in the region were hand dug and thus equipment failure was just one aspect of the risk faced by pastoralists. Lee (1984) noted that Herero occupation of wells in the Dobe area caused the water table to be lowered dramatically. High utilisation, particularly in the dry season when the water table is at its lowest, can cause wells to dry out and the only solution is to try and dig deeper or move to alternative water sources. The life span and cost of a well depends heavily on whether it is dug through hard calcrete or soft sand. Another serious risk is that wells dug in areas with sandy substrata are prone to collapse (Figure 3)

Figure 3: A well at the Kwarongo cattlepost. (a) Under construction, photo taken in October 2005 (top); and (b) 6 months after it collapsed, photo taken in June 2006 (below). Source: fieldwork.

Since receiving their new livestock, many pastoralists with large herds have adapted to the increased risk through dual well ownership and herd splitting. A number of these households acquired second wells when several became vacant during the Herero repatriation. Most of those who remain reliant on one well spoke of plans or pending applications for a second well permit from the Land Board. These strategies appear a direct adaptation to the increased risks involved as a result of enclosure.

One household fortunate enough to secure a second well was SM's family. Two brothers, MM and SM, cooperatively manage their large herd of cattle between two wells Marunga and Ndivitama. The family's main homestead (onganda) was originally located at Xaudum from where they would occasionally make a seasonal cattle camp at the Marunga waterhole. During the 1980s, the family dug a permanent well at Marunga after the water table dropped at Xaudum and the area became overgrazed. They continued to move seasonally between the two wells until the cull. After being restocked at Chukumuchu, however, they were told by DAHP officials to remain within the veterinary zone.

According to MM, the well at Marunga is one of the deepest in the area and requires a Lister engine to operate. These are more expensive to purchase and maintain. To mitigate the risk of Marunga's well failing the family purchased the Ndivitama well from its Herero owners when they repatriated in 1994. MM explained how the brothers cooperatively manage their herd between the two wells:

'We split our herd between Ndivitama and Marunga because of water. We keep more cows at Marunga as there is more water in that well and it doesn't dry out easily like Ndivitama. There are many waterholes near Marunga also. In the wet season we even take cattle from Ndivitama to Marunga. We then herd them towards the waterholes and fetch them again in the evening.'(October 2005)

The way SM's family adapted their herd management system illustrates both the gradual process of sedentarisation that has occurred in the region and the role played by veterinary cordon fence enclosure. It also shows that mobility and flexibility remain important strategies through which pastoralists manage enclosure and sedentarisation. Whereas, in the past the family managed their livestock between the grasslands of the Xaudum valley and Marunga, they now maintain mobility between Marunga and Ndivitama. Dual well ownership and herd splitting represent important strategies by which pastoralists manage the risks and vulnerability of enclosure.

Another major risk facing people in the region is fire. Enclosure by veterinary cordon fences in the region has increased the vulnerability of pastoralists to grazing shortages following periodic fire events. WK a 32-year-old Herero man from Kawaronga cattlepost explained why he resented the CBPP fences:

'Look at the land now, it's all burnt up. Before the CBPP cull I could have gone in search of grass for my cattle where the fire hasn't burnt and look for water and then returned when the rains arrived and my area was green. We don't need this fence, before we could trek cattle easily to Shakawe to the community butchers, now we need permits. The fence even kills wild animals (November 2005)'

Historically, pastoralists use fire as a management tool to encourage the growth of grasses and control woody plants and pests (McGahey et al. 2007). In the Kalahari, hunter-gatherer groups have traditionally burnt the savanah to remove dead vegetation and to facilitate hunting and gathering (Perkins et al. 2002). The savannah fire problem even concerned pastoralists who were in favour of enclosure due to the improved security situation. After explaining his overall positive position on fencing MS, a Mbukushu man from Chukumuchu admitted there are problems,

'When the fire burns and then there are poor rains, there is little grazing left for cattle between the fences, but there may be grazing to the north or south in these circumstances (July 2006)'

The issue is compounded by a dramatic increase in the spatial distribution and temporal frequency of burn mosaics throughout the Kalahari since the droughts of the 1980s (Perkins et al. 2002). Most people have noticed a similar trend in the Chukumuchu area. MK, a 52-year-old Mbukushu man explained:

'These days fires are a problem, they burn all over our area. During the 1980 s you couldn't see fires that often, they would burn only here or there. Even in the early 1990 s when Chukumuchu School was built fires were not this bad. It's only since the cull that fires have become this bad. The problem is this conservation of the environment idea (October 2005)'.

According to Perkins et al. (2002) there are two possible reasons why fires in the Kalahari have become more severe and widespread. First, the almost unprecedented decline in large wild herbivore numbers has increased the volume of dead biomass. Almost every person interviewed from the area stated that there has been a decline in wild herbivores, particularly since the CBPP fences were constructed. This is unsurprising given that veterinary cordon fences represent impermeable barriers for most species and are associated with large-scale declines in water-dependent species in the Kalahari (Thouless 1998, Wheelwright et al. 1996).

The second reason is what MK refers to as the 'conservation of the environment' idea. This concerns the widespread ban on traditional burning practices in the Kalahari which, coupled with the sedentarisation and shift from hunter gathering to livestock-dominated modes of production, has caused a reduction in the diversity of spatial and temporal burn mosaics. This, in turn, has increased the frequency of intense and widespread fires (Perkins et al. 2002). The incidence of fires originating from close to fences is also reported to have risen following increased human activity along fence roads (Scott Wilson 2000). Finally, whilst veterinary fences can also act as fire breaks, they often trap wildlife attempting to escape bush fires (ibid: 10), with resultant population declines exacerbating the fuel load problem.

It is likely that the reduced stocking rates following the CBPP cull and Herero repatriation have caused a further increase in the volume of dead plant biomass and fire intensity. This was confirmed by local observations of recent environmental changes in the region. According to local people there was an increase of unpalatable shrubs around the cattleposts before the cull. After the livestock were culled, however, the encroachment became even worse as cattle trampling had kept these areas relatively contained.

Whilst shrub encroachment may have increased close to kraals and wells, these areas are found in valley floors where savannah fires often burn out due to the lack of grass. On the dune top savannah, however, intense savannah fires have caused a reduction in mature trees and shrubs. According to natural resource gatherers, savannah fires now regularly destroy crops of wild berries and thatching grass, whilst the reduction in mature trees has reduced the bee population making honey hunting more difficult.

Restricting Resource Access: Young Pastoral Households Under 40 yrs Old

The final group of herders who hold negative positions on the cordon fences were predominantly young pastoral households. Although many young people still plan to farm for cultural reasons, these days most young people prioritise formal and informal employment opportunities as their preferred livelihood activity. During a meeting with a group of young teenagers in Chukumuchu, they discussed the pros and cons of farming and the role it may play in their future livelihoods:

SJ (male 27 years): It's difficult to keep cattle these days. Fires burn all the grass and you may need to buy feed for the cattle. Raising buckets all day to water your cows is hard work so you have to buy a water tank and generator. These things are expensive but its easy to solve these problems if you have a job you can buy fuel and pumps. Without these things you are in trouble. I want to continue making bricks and working as a builder. Maybe one day I will farm, its Tswana culture (October 2005).

KG (female 19 years): I don't want to keep fields, its hard work ploughing and harvesting. Then you have to pound the crop and fix the fences. I want to go to technical college and learn textiles (October 2005).

Whilst there may be an abundance of young people seeking formal and informal cash income opportunities in villages and towns, there were several young households, mainly in their thirties, attempting to establish themselves as independent pastoralists. For the Herero it is traditional for sons, especially the youngest, to 'break with their father' upon reaching adulthood and seek to establish their own cattlepost (Twyman et al. 2001). In the Chukumuchu region, there were several young Hambukushu households attempting to become autonomous as well, though this may be more out of necessity than tradition. These households were often resource poor, with no village-based homestead or well permit. The problems faced by this group in their efforts to gain access to the resources required to construct a pastoral livelihood are demonstrated by the case of MJ and his family.

MJ is 39 years old and was born at Nxau Nxau a village just south of the Ikoga fence where most of his relatives still live. Before the CBPP cull in 1996, MJ herded his 25 cows together with his father's cattle, but like many the family decided to opt for 100% cash compensation amid speculation that cattle would not be replaced by the government. Since this time, MJ has been busy building his own herd, primarily by working for the Herero in return for access to water and grazing. Households such as MJ's often lack the financial capital to invest in the expensive and risky process of well construction. As a result they find it hard to secure access to permanent water sources that are reliable, free and close to sufficient grazing resources. These issues were central to the negative fence opinion of this group of pastoralists as MJ explained,

'Before the fence I could have taken my cattle to Nxau Nxau where there are lots of wells that I can use for free, now I cannot go through the fence and have to either pay to use a well or work for the well owner (July 2006)'.

For households such as MJ's, there are few options open to access both adequate water and grazing. The only free water available in the Chukumuchu veterinary zone is a government borehole within the heavily grazed communal rangelands surrounding the village of Chukumuchu (Figure 4). However, in addition to the limited availability of grazing, keeping large herds of livestock close to villages is likely to end in conflict as livestock damage crops and other property. Whilst constructing a deep well that may be beyond most pastoralists in this group, there are areas where water can be accessed within a short distance from the surface. Most of these, however, are found along the Xaudum valley to the north or near Nxau Nxau village to the south. The only place where shallow wells can be dug in the Chukumuchu region is at Tsokung, a cattlepost already occupied by several households.

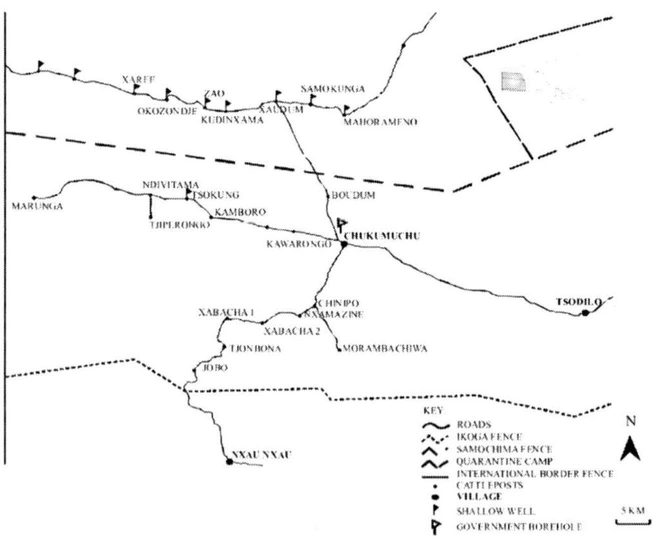

Figure 4: Shallow wells and government boreholes in the Chukumuchu region. (NB: water sources south of Ikoga fence not included in survey). Source: fieldwork.

There are other key factors besides access to adequate water which such households must consider when attempting to become independent pastoral households. For young herding households such as MJ, who sought access to wells near Nxau Nxau, proximity to villages is also a major concern, as these groups are more reliant on income generated from trading natural resource-based goods to village dwellers,

'I want the Ikoga fence to be removed because there is not enough grazing or water this side for us. If the fence was not there now I would be able to go to Nxau Nxau and find water as there are plenty of waterholes and wells I can use there. Now I have to come up here and beg these Herero to use their well. I would rather be closer to Nxau Nxau as my wife can sell milk and oil in the village and buy maize meal for the family (July 2006)'.

Young families such as MJ's are finding it increasingly difficult to establish themselves as independent herding households. Constrained between the need to access adequate grazing and water on the one hand, and settle within easy distance of settlements in order to participate in local markets on the other, they have been further disadvantaged by the CBPP fences. Such households are often forced to move from one cattlepost to another in search of adequate water and grazing. Livestock are important assets from which the rural poor can construct routes out of poverty, and whilst poorer households may be more reliant on small stock and forage close to settlements, there are examples where some households have managed to accumulate stock and access to extensive rangelands are critical for this to occur (Toulmin et al. 2004, Ellis and Mdoe 2003). It would appear, therefore, that enclosure in the context of the CBPP fences could act to enhance social differentiation within the region's pastoral society.

CONCLUSIONS: WIDER SIGNIFICANCE FOR PASTORAL SYSTEMS IN AFRICA

Veterinary cordon fences represent the lynchpin of the southern African disease control system and in Botswana they enable livestock movement control upon otherwise open-access communal rangeland. The continued expansion of veterinary cordon fences into northern regions has facilitated secured access to lucrative export markets for commercial livestock farmers throughout the south of the country. These fences and the area-based disease freedom policies that surround their use represent a striking barrier to pastoral mobility, yet so far few studies have investigated the impact of these structures upon the subsistence pastoral economy.

Tracking changes in herd management, in particular declining livestock mobility, over the period leading up to and including veterinary cordon fence enclosure proved vital to contextualise perceived impacts emerging from the pastoral community. These enquiries revealed that three overall seasonal herd mobility strategies were employed in the past: sedentary herd-release management, wet seasonal mobility in which herders moved to a single water source to avoid conflicts on arable lands, and wet season mobility to multiple water sources to track resource availability. Several authors have stressed the importance of socio-economic factors over ecological considerations in determining herding decisions within contemporary pastoral communities (cf. Baker and Hoffman 2006, Turner 1999). Socio-economic changes associated with a gradual shift towards wage-dependency and regional economic development reduced labour availability at the cattleposts, which in turn has led to a reduction in seasonal herd mobility. This empirical evidence demonstrates exactly how household herd management in pastoral societies can become increasingly dominated by non-environmental considerations with increased income diversification.

One might assume that a colonial-era policy instrument that focuses on containment and movement restriction would impact negatively upon a subsistence pastoral society whose primary focus may well be the maintenance of opportunistic and flexible rangeland access (Scoones 1995). Using a detailed examination of the precise societal costs associated with two cordon fences in northern Botswana, this paper illustrates that the situation is less than straightforward. In fact, the majority of herders welcome enclosure, especially those faced with labour shortages after the general movement towards non-agricultural forms of employment. For these groups, veterinary cordon fences reduce the risk of stock theft and provide a useful limit to the movement of lost livestock, especially during the wet season. Furthermore, the increased presence of veterinary personnel either at sporadic gate camps or on regular maintenance patrols represent additional resources that pastoralists can draw upon to locate lost livestock. In Namibia, pastoralists have intentionally created fenced community enclosures, primarily to reduce herding requirements and improve herd security (Werner 2000, Stahl 2000, Twyman et al. 2001). It would appear that faced with declining agricultural labour availability, herders throughout the region are beginning to favour fencing and enclosure over more opportunistic and flexible resource access arrangements.

Whilst the maintenance of flexible resource access may have become of lesser importance to many herders, a small minority of pastoralists outwardly resent the fences (notably elderly herd patriarchs and young herding households). Difficulties experienced gaining access to key resources is creating resentment towards the fences from young pastoral households attempting to 'break with their father' and establish their own household. Constrained between the need to access equitable grazing and water resources on the one hand, and settle within easy distance of settlements in order to participate in local markets on the other, they have been further disadvantaged by the fences. Further work would be needed to determine the wider extent of the problem. This is important, given that studies investigating fencing and enclosure typically emphasise that the removal of flexible systems of resource access disadvantages marginalised groups creating greater social differentiation (Taylor 2006; Peters 1994).

The importance of access qualifications in determining the ability of people to engage their optimal risk management strategies is also emphasised by analysts of societal vulnerability to natural hazards (Wisner et al. 2004). Well failure, savannah fires and drought represent persistent risks or hazards to which livestock production in the region is exposed. Awareness of the increased vulnerability to hazards such as savannah fires and hand-dug well failure that enclosure presents caused some experienced pastoralists to oppose enclosure. In a direct attempt to mitigate their increased vulnerability to natural hazards, these herders have attempted to adapt to enclosure by securing dual well ownership, herd splitting and practising seasonal mobility. Further work is needed to investigate how pastoralists enclosed by veterinary fences cope with risks over longer time periods. This temporal approach is especially important given that long-term exposure to hazards can initiate a 'ratchet effect' whereby the ability of households to respond to subsequent shocks may be lessened (Wisner et al. 2004).

In extending marketing opportunities to marginalised pastoral areas, disease management in Botswana is often regarded as a model that could help to catalyse Africa's currently underdeveloped potential for livestock production. Such assertions surround current efforts to re-establish a fenced export zone in eastern Africa, for example (Kock 2010). Focusing on the case of veterinary cordon fencing in northern Botswana, this paper illustrates the exact societal costs and benefits associated with the policy instrument. Whilst on balance fences may have reduced herding demands and improved livestock security, negative implications for resource access and pastoral risk management raise concern.

From a wider policy perspective, questioning these assumptions is timely given the recent animal health policy dilemmas facing the southern African region (cf. Scoones and Wolmer 2006, Scoones et al. 2010). With rapidly changing political, economic and policy contexts, it is argued that the conventional area-based animal health policy is unlikely to remain relevant in its current form in southern Africa, especially with uncertainty over the future of preferential trade agreements and escalating costs associated with sanitary and phyto-sanitary compliance. This is an increasingly complex issue with a range of scientifically acceptable options of disease control and livestock marketing opening up before decision makers (for a review see Scoones et al. 2010). The findings emerging from this study suggest that policy-marketing pathways which require less fencing and offer more flexible resource access such as a commodity-based trade (Thomson et al. 2004), may well be preferable within communal pastoral areas.

AUTHORS' CONTRIBUTIONS

DJM was solely responsible for the preparation of material presented in this manuscript including all data collection, analysis and write up.

ACKNOWLEDGEMENTS

Research referred to throughout this paper was funded by the Economic and Social Research Council, Natural Environmental Research Council and Hertford College Oxford. My thanks are extended to the people of

Chukumuchu for taking the time to participate in the research and to the Department of Animal Health and Production for granting a permit to conduct this study. I also extend my gratitude to David Thomas and Patricia Daley of Oxford University for their valuable support, advice and comments on this research. Similarly, thanks to the referees of this paper for their valuable comments.

REFERENCES

1. Adriansen, HK, and TT Nielsen. 2002. Going where the grass is greener: On the study of pastoral mobility in Ferlo, Senegal. Human Ecology 30: 215–226.
2. Almagor, U. 1981. Pastoral identity and reluctance to change: The Mbanderu of Ngamiland. Journal of African Law 24: 35–61.
3. Baker, LE, and MT Hoffman. 2006. Managing variability: Herding strategies in communal rangelands of semiarid Namaqualand, South Africa. Human Ecology 34: 765–784.
4. Behnke, RH, I Scoones, and C Kerven. 1993. Range ecology at disequilibrium: New models of natural variability and pastoral adaptation in African savannas. 248. London: ODI.
5. Bensen, H, and T Meyer. 2002. The dynamics of land use systems in Ngamiland: Changing livelihood options and strategies. [http://www.okavangochallenge.com] Accessed 15 January 2005
6. Boone, RB, and NT Hobbs. 2003. Lines around fragments: The effects of fencing on large herbivores. ed. Allsopp, N, Palmer, A R, Milton, S J, Kirkman, K P, Kerley, G I H, Hurt, C R and Brown, C J, 580–591. Proceedings of the VIIth International Rangelands Congress, Durban, South Africa.
7. Bollig, M, and JB Gewald. 2000. People, cattle and land: Transformations of a pastoral society in southwestern. Africa, Koln: Rudiger Koppe Verlag.
8. Coppolillo, PB. 2000. The landscape ecology of pastoral herding: Spatial analysis of land use and livestock production in East Africa. Human Ecology 28: 527–560.
9. Darkoh, MBK, and JE Mbaiwa. 2002. Globalisation and the livestock industry in Botswana. Singapore Journal of Tropical Geography 23 (2): 149–166.

10. Davies, J, and R Hatfield. 2007. The economics of mobile pastoralism: A global summary. Nomadic Peoples 11 (1): 91–116.
11. Desta, S, and DL Coppock. 2004. Pastoralism under pressure: Tracking system change in southern Ethiopia. Human Ecology 32 (4): 465–486.
12. Ellis, F. 2000. Rural livelihoods and diversity in developing countries. Oxford: Oxford University Press.
13. Ellis, F, and N Mdoe. 2003. Livelihoods and rural poverty reduction in Tanzania. World Development 31: 1367–1384.
14. Fidzani, B, WS Mlenga, M Atlhopheng, and MM Shatera. 1999. Socio-economic effects of CBPP in Ngamiland, Division of Agricultural Planning and Statistics, MoA.
15. Fratkin, E. 1997. Pastoralism: Governance and development issues. Annual Review of Anthropology 26: 235–261.
16. Fratkin, E, and R Mearns. 2003. Sustainability and pastoral livelihoods: Lessons from east African Maasai and Mongolia. Human Organization 62: 112–122.
17. Hitchcock, RK. 1995. Centralisation, resource depletion, and coercive conservation amoung the Tyua of the Northern Kalahari. Human Ecology 23 (2): 169–196.
18. Hitchcock, R. 2002. Coping with uncertainty: Adaptive responses to drought and livestock disease in the northern Kalahari. In Sustainable Livelihoods in Kalahari Environments: Contributions to Global Debates, ed. Sporton, D, Thomas, DSG, 161–192. Oxford: Oxford University Press.
19. Hobbs, JCA. 1981. The environmental impact of veterinary cordon fences. African Wildlife 35 (6): 16–21.
20. Jerve, AM. 1982. Cattle and inequality: A study of rural differentiation from southern Kgalagadi in Botswana. Bergen: DERAP Publications, Chr. Michelson Institute.
21. Kivaria, FM. 2003. Foot and mouth disease in Tanzania: An overview of its national status. Veterinary Quarterly 25: 72–78.
22. Kruger, FW. 1998. Identity building and social transformation: The cases of Namibia and Botswana compared. Geojournal 46 (2): 79–87.

23. Kock, R. 2010. The newly proposed Laikipia disease control fence in Kenya. In Fencing impacts: A review of the environmental, social and economic impacts of game and veterinary fencing in Africa with particular reference to the Great Limpopo and Kavango-Zambezi Transfronier Conservation Areas, ed. Ferguson, K, Hanks, J, 71–74. Pretoria: University of Pretoria, Mammal Research Institute.

24. Lee, R. 1984. The Dobe Ju/'hoansi. Toronto: Holt, Rinehart and Winston, Inc.

25. Madsen, LM, and HK Adriansen. 2004. Understanding the use of rural space: The need for multi-methods. Journal of Rural Studies 20: 485–497.

26. Manzano, P, and R Casas. 2010. Past, present and future of Transhumancia in Spain: Nomadism in a developed country. Pastoralism 1 (1): 72–90.

27. McAllister, RRJ. 2010. Livestock mobility in arid and semi-arid Australia: Escaping variability in space. Pastoralism 1 (1): 37–54.

28. McGahey, D, J Davies, and E Barrow. 2007. Pastoralism as Conservation in the Horn of Africa: Effective policies for conservation outcomes in the drylands of eastern Africa. Annals of the Arid Zone 46: 353–377.

29. McPeak, JG, and PD Little. 2006. Pastoral Livestock Marketing in Eastern Africa: Research and Policy Challenges. 288. London: Intermediate Technology Publications Ltd.

30. Moleele, NM, and JS Perkins. 1998. Encroaching woody plant species and boreholes: Is cattle density the main driving factor in the Olifants Drift communal grazing lands, south-eastern Botswana? Journal of Arid Environments 40 (3): 245–253.

31. Moore, DS. 1993. Contesting terrain in Zimbabwe's eastern highlands: Political ecology, ethnography and peasant resource struggles. Economic Geography 69: 380–401.

32. Mullins, GR, B Fidzani, and M Kolanyane. 2000. At the end of the day--the socioeconomic impacts of eradicating contagious bovine pleuropneumonia from Botswana. Tropical Veterinary Diseases 916: 333–344.

33. Ndzinge, L, J Marsh, and R Greer. 1984. Herd inventory and slaughter supply response of Botswana beef cattle producers. Journal of Agricultural Economics 35: 97–107.

34. Niamir-Fuller, M, and MT Turner. 1999. A review of recent literature on pastoralism and transhumance in Africa. In Managing mobility in African rangelands: The legitimization of transhumance, ed. Niamir-Fuller, M, 18–46. London IT Publications.
35. Niamir-Fuller, M(ed). 1999. Managing mobility in African rangelands: The legitimization of transhumance. London: IT Publications.
36. Pennington, R. 1992. Did food increase fertility? Evaluation of !Kung and Herero History. Human Biology 64: 497.
37. Pennington, R, and H Harpending. 1991. Infertility in Herero pastoralists of Southern Africa. American Journal of Human Biology 3: 135–153.
38. Perkins, JS, G Stuart-Hill, and B Kgabung. 2002. The impact of cattle-keeping on the wildlife, vegetation, and veld products. In Sustainable Livelihoods in Kalahari Environments:Contributions to Global Debates, ed. Sporton, D, Thomas, DSG, 67–90. Oxford: Oxford University Press.
39. Perkins, JS. 1996. Botswana: Fencing out the equity issue. Cattleposts and cattle ranching in the Kalahari Desert. Journal of Arid Environments 33: 503–517.
40. Perkins, JS, and DSG Thomas. 1993. Spreading deserts or spatially confined environmental impacts? Land degradation and cattle ranching in the Kalahari of Botswana. Land Degradation and Rehabilitation 4: 179–194.
41. Peters, P. 1994. Dividing the Commons: Politics, Policy, and Culture in Botswana. Charlottesville: University Press of Virginia.
42. Reid, RS, PK Thornton, and RL Kruska. 2003. Loss and fragmentation of habitat for pastoral people and wildlife in east Africa: Concepts and issues. In Proceedings of the VIIth International Rangelands Congress, Durban, South Africa, ed. Allsopp, N, Palmer, AR, Milton, SJ, Kirkman, KP, Kerley, GIH, Hurt, CR, Brown, CJ, 580–591.
43. Republic of Botswana. 2001. 2001 Population and housing census: Population of towns, villages and associated localities. 167. Gaborone: Central Statistics Office, Government Printer.
44. Safriel, U, Z Adeel, D Niemeijer, J Puigdefabregas, R White, R Lal, M Winslow, J Ziedler, S Prince, E Archer, and C King.

2005. Chapter 22: Dryland systems. In Millenium Ecosystem Assessment. In Ecosystems and human well-being: Current state and trends, ed. Hassan, R, Scholes, R, Ash, N, 1: 623–662. Washington, DC: World Resources Institute.

45. Scoones, I. 1995. Living with uncertainty: New directions for pastoral development in Africa. London: Intermediate Technology Publications.

46. Scoones, I, and W Wolmer. 2006. Livestock, disease, trade and markets: Policy choices for the livestock sector in africa. 269. Brighton: IDS Working Paper.

47. Scoones, I, A Bishi, N Mapitse, R Moerane, ML Penrith, R Sibanda, G Thomson, and W Wolmer. 2010. Foot-and-mouth disease and market access: Challenges for the beef industry in southern Africa. Pastoralism 1 (2): 135–164.

48. Scott Wilson Resource Consultants. 2000. Environmental impact assessment of the veterinary fences in Ngamiland. Summary report for the Government of Botswana Edinburgh. 105.

49. Smith, AB. 1992. Pastoralism in Africa: Origins and development ecology. London: Hurst and Company.

50. Stahl, U. 2000. At the end of the day we will fight: Communal land rights and illegal fencing in the Otjozondjupa Region. In People, Cattle and Land: Transformations of a Pastoral Society in Southwestern Africa, ed. Bollig, M, Gewald, J, 319–346. Koln: Rudiger Koppe Verlag.

51. Sutmoller, P. 2002. The fencing issue relative to the control of foot-and-mouth disease. Annals of the New York Academy of Sciences 969: 191–200.

52. Taylor, JL. 2006. Negotiating the Grassland: The Policy of Pasture Enclosures and Contested Resource Use in Inner Mongolia. Human Organization 65 (4): 374–386.

53. Theis, J, and H Grady. 1991. Participatory Rapid Appraisal for Community Development: A training manual based on experiences in the Middle East and North Africa. London: IIED.

54. Thomas, DSG, and PA Shaw. 1991. The Kalahari Environment. Cambridge: Cambridge University Press.

55. Thomson, G, E Tambi, S Hargreaves, T Leyland, A Catley, G van 't Klooster, and M Penrith. 2004. International trade in livestock and

livestock products: The need for a commodity-based approach. Veterinary Record 155: 429–33. PubMed Abstract

56. Thouless, C. 1998. Large mammals inside and outside protected areas in the Kalahari. Transactions of the Royal Society of South Africa 53: 245–255.
57. Toulmin, C, C Hesse, and L Cotula. 2004. Pastoral common sense: Lessons from recent developments in policy, law and practice for the management of grazing lands. Forests, Trees and Livelihoods 14: 243–262.
58. Turner, MD. 1999. Labor process and the environment: The effects of labor availability and compensation on the quality of herding in the Sahel. Human Ecology 27: 267–296.
59. Turner, MD, and P Hiernaux. 2002. The use of herders' accounts to map livestock activities across agropastoral landscapes in semi-arid Africa. Landscape Ecology 17: 367–385.
60. Twyman, C, A Dougill, D Sporton, and D Thomas. 2001. Community fencing in open rangelands: Self-empowerment in Eastern Namibia. Review of African Political Economy 87: 9–26.
61. UNDP. 2003. Pastoralism and mobility in the drylands. The global drylands imperative, Second Challenge Paper Series. [http://www.energyandenvironment.undp.org/undp/index.cfm?module=Library&page=Document&DocumentID = 5572]Accessed on 5 January 2005
62. Vivelo, FR. 1977. The Herero of western Botswana: Aspects of change in a group of Bantu-speaking cattle herders. 232. St. Paul: West Publishing Co.
63. Werner, W. 2000. From communal pastures to enclosures: The development of land tenure in Herero reserves. In People, Cattle and Land: Transformations of a Pastoral Society in Southwestern Africa, ed. Bollig, M and J Gewald, Koln: Rudiger Koppe Verlag.
64. Wheelwright, NT, JM Dangerfield, M Flyman, and M Tjibae. 1996. Strengthening biological theory in wildlife research and management: A case study for Botswana. South African Journal of Science 92 (4): 173–177.
65. Wisner, B, P Blaikie, T Cannon, and I Davies. 2004. At risk: Natural hazards, people's vulnerability and disasters. London: Routledge.15-24-6.

Citations

CHAPTER 1

David C. Hall and Quynh Ba Le, "A Basic Strategy to Manage Global Health with Reference to Livestock Production in Asia," Veterinary Medicine International, vol. 2011, Article ID 328307, 6 pages, 2011. doi:10.4061/2011/328307.

CHAPTER 2

Ayodele O Majekodunmi, Akinyemi Fajinmi, Charles Dongkum, Alexandra P M Shaw, and Susan C Welburn, Pastoral livelihoods of the Fulani on the Jos Plateau of Nigeria, doi:10.1186/s13570-014-0020-7

CHAPTER 3

Thomas C Pierson, Nathan J Wood, and Carolyn L Driedger, Reducing Risk From Lahar Hazards: Concepts, Case Studies, and Roles for Scientists, doi:10.1186/s13617-014-0016-4.

CHAPTER 4

Francis EO Opiyo, Oliver V Wasonga, and Moses M Nyangito, Measuring household vulnerability to climate-induced stresses in pastoral rangelands of Kenya: Implications for resilience programming, doi: 10.1186/s13570-014-0010-9.

CHAPTER 5

Rupak Goswami, Soumitra Chatterjee, and Binoy Prasad, Farm Types and Their Economic Characterization in Complex agro-Ecosystems for Informed Extension Intervention: Study from Coastal West Bengal, India, doi:10.1186/s40100-014-0005-2.

CHAPTER 6

Diphetogo Mosalagae and Kabo Mogotsi, Caught in a Sandstorm: an Assessment of Pressures on Communal Pastoral Livelihoods in The Kalahari Desert of Botswana, doi:10.1186/2041-7136-3-18.

CHAPTER 7

Fred Nelson, Natural Conservationists? Evaluating the Impact of Pastoralist Land Use Practices on Tanzania's Wildlife Economy, doi:10.1186/2041-7136-2-15.

CHAPTER 8

Daniel J McGahey, Livestock Mobility and Animal Health Policy in Southern Africa: the Impact of Veterinary Cordon Fences on Pastoralists, doi:10.1186/2041-7136-1-14.

Index